AUTISM SPECTRUM DISORDERS: THE ROLE OF GENETICS IN DIAGNOSIS AND TREATMENT

Edited by **Stephen I. Deutsch** and **Maria R. Urbano**

Autism Spectrum Disorders: The Role of Genetics in Diagnosis and Treatment
Edited by Stephen I. Deutsch and Maria R. Urbano

Published by InTech
Janeza Trdine 9, 51000 Rijeka, Croatia

Publishing Process Manager Ivana Lorkovic

Technical Editor Teodora Smiljanic

Cover Designer InTech Design Team

Additional hard copies can be obtained from orders@intechopen.com

Autism Spectrum Disorders: The Role of Genetics in Diagnosis and Treatment, Edited by Stephen I. Deutsch and Maria R. Urbano
p. cm.
ISBN 978-953-307-495-5

For EU product safety concerns:
IN TECH d.o.o., Prolaz Marije Krucifikse Kozulić 3, 51000 Rijeka, Croatia,
info@intechopen.com or visit our website at intechopen.com.

Meet the editors

Dr. Stephen I. Deutsch received his medical and graduate degrees in biochemical pharmacology from the New York University School of Medicine. He conducted his graduate research in the laboratory of Dr. Rody P. Cox, working on the molecular mechanisms of nonketotic hyperglycemia, which provided early insights into glycine's neurotransmitter and metabolic roles in the CNS, and effects of sodium butyrate on changes in morphology, rates of cell division and gene expression in HeLa cells. Dr. Deutsch completed training in Psychiatry at the NYU-Bellevue Psychiatric Hospital, where he worked with Drs. Samuel Gershon and Magda Campbell on clinical trials and development of rational pharmacotherapies for major psychiatric disorders. Dr. Deutsch completed a Pharmacology Research fellowship at the NIMH, working in the laboratory of Dr. Steven M. Paul on the in vivo characterization of the central benzodiazepine binding site. Dr. Deutsch has published over 240 original articles, reviews and book chapters. Dr. Deutsch was a Distinguished NARSAD Investigator and is a Fellow of the American College of Neuropsychopharmacology.

Dr. Maria R. Urbano received her medical degree from Eastern Virginia Medical School after completing a Master of Science degree in microbiology at Michigan State University. She completed an internship in internal medicine and then completed her residency in psychiatry at Eastern Virginia Medical School. She is board certified in general psychiatry. She is a Fellow of the American Psychiatric Association and an honorary member of the Alpha Omega Alpha Honor Society. She received a Master Clinician Award in psychopharmacology from the Neuroscience Education Institute. As an Associate Professor of the Dept. of Psychiatry and Behavioral Sciences, Dr. Urbano has been very active in the education of both medical students and psychiatry residents. She has twenty five years of clinical psychiatry experience and is currently the Clinical Director of the Autism Spectrum Disorders Program at Eastern Virginia Medical School. As clinical director of the ASD Program, she is currently involved in several translational research studies in collaboration with Dr. Stephen I Deutsch, MD, PhD, et al Studies clarifying comorbid psychiatric diagnoses that may accompany ASD are also underway.

Contents

Preface

The broadening of the definitional criteria of autism spectrum disorders (ASDs) and increased recognition of these syndromes have led to dramatic increases in their estimated prevalence; prevalence estimates of ASDs in the USA are approximately 1 in 110 children with a three to four time greater male to female predominance. These disorders occur commonly as co-morbid conditions in several Mendelian genetic disorders due to the effects of a single major gene (e.g., tuberous sclerosis). Importantly, although these Mendelian disorders appear to be unrelated to each other, recent advances in bioinformatics and "network analyses" suggest that they may indeed be related to each other; the points of convergence can include development and architecture of the synapse, and early developmental events in neurogenesis, neuronal cell migration and synaptogenesis. Additionally, areas along the human genome are emerging as "hotspots" for microdeletions and microduplications, referred to as Copy Number Variants (CNVs); the density of these CNVs may contribute to increased risk of neurodevelopmental syndromes, including ASDs. Remarkably, although the 1970's was focused on elucidating descriptive differences between ASDs and schizophrenia presenting in childhood; the emerging data on CNVs suggest that ASDs and schizophrenia, or at least their genetic mechanisms, may be more similar than initially appreciated. In any event, the genetic data are also suggesting molecular targets; for example, microdeletions at the 15q13.3 locus suggest that haploinsufficiency of a gene product of this locus (i.e., CHRNA7), which codes for the $\alpha 7$ nicotinic acetylcholine receptor ($\alpha 7$ nAChR) subunit, may be causally associated with ASDs. Thus, selective nicotinic acetylcholine receptor agonist strategies should be explored for their potential therapeutic benefit. The high prevalence of these disorders, their impact on the identified affected patient and the unrecognized unaffected family members (including sibs), accessibility of Array Comparative Genomic Hybridization screening technologies, elucidation of associations with candidate susceptibility genes, along with CNVs and complex genetics are raising profound ethical questions, heightening the challenges of genetic counseling. The staggering challenges of genetic counseling are further compounded by issues of imprinting (i.e., homologous maternal and paternal chromosomes may have different patterns of cytosine methylations and certain genetic disorders differ depending on genetic variations within one of the affected parental chromosomes [e.g., Angelman and

Prader-Willi syndromes]) and variable "penetrance" (i.e., there is a broad array of possible phenotypes). The chapters contained in this book highlight some of these emerging issues.

Stephen I. Deutsch, M.D., Ph.D. and Maria R. Urbano, M.D.
Department of Psychiatry and Behavioral Sciences
Eastern Virginia Medical School
825 Fairfax Avenue, Suite 710
Norfolk, Virginia 23507-1912
USA

Part 1

Early Recognition and Diagnosis

1

Early Detection of Autism Spectrum Disorders

Jariya Chuthapisith and Nichara Ruangdaraganon
Department of Paediatrics, Faculty of Medicine Ramathibodi Hospital
Mahidol University, Bangkok
Thailand

1. Introduction

Autism spectrum disorders (ASDs) are neurodevelopmental disorders characterized by distinctive language impairments, social and communicative deficits, and patterns of restricted and stereotyped behavior. In the Diagnostic and Statistical Manual of Mental Disorders, fourth edition, text revision (DSM-IV-TR) (American Psychiatric Association, 2000), pervasive developmental disorders (PDDs) are also referred to as autistic disorder (AD), Asperger's disorder, PDD not otherwise specified (PDD-NOS), childhood disintegrative disorder, and Rett Disorder. However, the diagnostic boundaries between these PDD subtypes remain unclear, the symptoms and behaviours lie on a continuum and have considerable clinical heterogeneity (Szatmari, 1999). In this review, therefore, ASDs are referred to as the diagnostic category of PDDs.

2. Diagnosis of ASDs

The manifestations of ASDs vary from mild to severe and pervasive impairment. Currently, the diagnosis of ASDs is based on the criteria developed in the DSM-IV-TR and the International Classification of Diseases, 10th revision (ICD-10) (World Health Organization (WHO), 1992) and is supported by standardized diagnostic instruments. According to the DSM-IV-TR criteria, the impairments of ASDs consist of three main impairments which must all be presented for diagnosis.

2.1 Impairment in social interaction is defined by various symptoms including impairment in the use of nonverbal behaviours (e.g. eye contact, use of gestures and facial expressions); lack of showing, bringing or pointing out objects; odd relationships of approaches to others; and lack of social or emotional reciprocity.

2.2 Impairments in communication consist of delay in or total lack of spoken language, inability to initiate or sustain a conversation with others, stereotyped or repetitive use of language, and lack of social imitative play.

2.3 Restricted repetitive and stereotyped patterns are behaviours, interests and activities as manifested by an inability to cope with change, a dislike for any interruption to routine, preoccupation with specific subjects or activities, repetitive or stereotyped motor mannerisms such as hand flapping or twisting, and persistent preoccupation with parts of objects.

Early diagnosis for ASDs is undoubtedly important and is considered as a clinical best practice. Early detection of ASDs leads to an early intervention (Rutter et al., 2006).

However, diagnosis before the age of 3 years remains a challenge (Baron-Cohen et al., 1996). Some symptoms of ASDs may overlap with normal developmental variance. Also, ASDs are a continuum of disease which has a wide range of individual differences. Distinctions between autistic disorder and PDD-NOS remain unstable. A study reported that up to 50% of PDD-NOS cases, who were diagnosed before age 3 years, could have been overdiagnosed, whereas around 22% were underdiagnosed (Chawarska et al., 2007). This was due to the fact that diagnosis depends on clinical judgments which sometimes may not agree with the DSM-IV-TR diagnostic criteria especially evaluating a young child. Some of the criteria in the DSM-IV-TR can not apply to young children. In other words, many of the characteristic behaviours in the DSM-IV-TR are not apparent before 36 months. For example, a child age less than 16-month-old typically can engage in parallel play but has not yet developed reciprocal peer relationships. Thus, the criteria of failure to develop age-appropriate peer relationships need to be adapted (Martinez-Pedraza & Carter, 2009). The criteria of stereotyped and repetitive use of language can be difficult to discriminate between repetitions of the last word in young typically developing children and echolalia in children with ASDs. Furthermore, the criteria "restricted repetitive and stereotyped patterns of behaviour, interests and activities" may not appear in young children. These may appear later after the third birthday in some cases (Gray & Tonge, 2001; Turner, 1999). Therefore, making a diagnosis in children younger than 2 years of age is very challenging.

3. Early signs of ASDs

Many research studies have concluded that the first signs and symptoms of ASDs are evident by 12 to 18 months of age (De Giacomo & Fombonne, 1998; Young et al., 2003). Research on early signs and symptoms of ASDs in young children have focused on parental retrospective reports, early home videos of children later diagnosed with ASDs, and studies on siblings of children with ASDs. The emergence of ASDs signs and symptoms involve the area of social skill deficits, language skill deficits and unusual repetitive or stereotypical behavioural patterns. Signs and symptoms that are predictive of ASDs in young children are, namely:

3.1 Social skills deficits

Social skills are one of the most important areas in defining ASDs in very young children. In typically developing children, social development is acquired parallel to overall development (e.g. language, motor and cognitive development). In the very young children whose language skills are limited, social development depends very much on clinical observations. The manifestation is a lack of or a decreased drive to connect with others, including share feelings, thoughts and actions. Children who have ASDs have limited or reduced eye contact, fail to orient their name being called, limited imitation, limited responding to reciprocal social games, and lack of showing or bringing an object to a caregiver.

The important characteristic in helping make a diagnosis in very young children is lack of "joint attention" (JA) (Charman, 2003; Dawson et al., 2002; Turner et al., 2006). JA refers to the capacity of the child to coordinate attention with a social partner in relation to an object or event (Rapin & Tuchman, 2008). JA normally appears to develop between 8-16 months. In 8-10 months old typically developing children, the child will follow the caregiver's gaze when the caregiver looks at an object or event. This development milestone is called "gaze monitoring". Around 10-12 months of age the child can follow the caregiver's point and can look back at the caregiver. At approximately 12-14 months the child will request for objects by pointing. In detail, the child will look back and forth between the object and caregiver to reassure that the caregiver understands his or her need, so called protoimperative pointing. At 14-16 months when the protodeclarative pointing develops, the child will look alternatively between the object and the caregiver. The goal is to share social experience, not the desired object (Johnson & Myers, 2007). Other nonverbal gestures, including facial expression, usually help discriminate the difference between these two types of pointing. Children with ASDs can not achieve these skills at an age-expected time or some can achieve partially but do not qualitatively achieve the skill completely. Some children may have no pointing at all but use their caregivers' hands point to the desired object. Some children look at the object but do not look at the caregiver to connect socially. A study in infant siblings of children with ASDs stated that the inability to shift one's attention (between child, parent and object) may be the first reliable sign of ASDs (Zwaigenbaum et al., 2005). In brief, lack of or delayed JA skill that is discrepant from overall functioning is a core feature of the ASDs diagnosis.

Since JA skills may not be observed in typically developing children younger than 1 year of age, responding to their name being called is a skill that the child should achieve. Children with ASDs usually fail to respond to their name being called. Some children with ASDs may respond to environmental sounds well enough to reassure the caregivers that their children can hear. Home videos of 1-year-old children who later were diagnosed with ASDs found that orienting to name being called is one of the most consistent deficits for affected children at that age (Baranek, 1999; Osterling & Dawson, 1994).

Delay in play skills is one of the features associated with diagnosis of ASDs. In respective order, play starts with sensory-motor, functional, constructive, and pretend or imaginary play. In typically developing children, approximately 4 months old, sensory-motor play begins. At 12-14 months of age, the child plays in a more functional manner. Pretend play starts around 16-18 months of age and increases gradually in complexity. Lack of or delay in pretend play or play that never passes the sensory-motor play stage serves as a distinguishing characteristic of ASDs. Although, some children with ASDs progress to functional play, the quality of play is significantly different from typically developing children by around age 2 years i.e. play is less purposeful, less symbolic and less in complexity (McDonough et al., 1997; Sigman et al., 1999; Stone et al., 1990). Some children with ASDs play or manipulate objects in a stereotypic or ritualistic manner such as lining up, banging, and mouthing objects. They usually prefer playing alone and have trouble incorporating into social play. This sophisticated social play may not develop which further worsen social skills development.

Although, there is a possibility to detect social skills deficits in children younger than 1 year of age, the reliability remain problematic before 18 months (Rutter, 2006). Special consideration should focus on gaze monitoring, joint attention, responding to being called by name, and play skills.

3.2 Early language skills deficits

Generally, absence of language skills appears at around age 2, which may lead to diagnosis of ASDs. In order to diagnose of ASDs earlier, delay in language development should be detected as soon as possible. A study among the siblings of children with ASDs demonstrated that during the first year of life, infants later diagnosed with autism vocalized less than low-risk control infants. Moreover, delays in verbal skills and early language comprehension were evident (Zwaigenbaum et al., 2005). Regarding language abnormalities, both expressive and receptive language deficits should be monitored. Typically, infants start to babble by 6 months of age, followed by advances in complexity which includes several phonemes. Later, jargoning (i.e. adds inflection to utterances in an attempt to tell a story) develops at approximately 10 -12 months of age. Lack or delay of an alternating to-and-fro pattern of vocalizations between infant and parent, delay of onset of babbling, and decrease or no use of pre-speech gestures (e.g. pointing, showing, nodding) are characteristic of ASDs (Wetherby et al., 2000; Johnson & Myers, 2007).

Repeating words in particular the last one or two words of a sentence right after being heard can be observed in typically developing children under the age of 2 years, which mimicks the ASDs symptom of immediate echolalia. However, the typically developing child will pass through this brief stage and will acquire functional language. In children with ASDs, this imitation still persists as expressive language after the age of around 2 years and beyond. Furthermore, the children with ASDs mostly repeat words in an odd intonation or repeat exactly the same intonation as they heard (Martinez-Pedraza & Carter, 2009).

In young children with ASDs, receptive language ability is often impaired. They initially do not respond to their names when called by a caregiver. After language is present, children with ASDs are unable to initiate or sustain conversation. Some children have comprehension deficits, particularly in complex sentences or questions. Children with ASDs also show deficits in non verbal communication; for example, they look at others less, have less social smile, lack appropriate gestures, have less pointing or have difficulty following a point, show objects less and have a lack of appropriate facial and emotional expression. These non verbal communication deficits are linked closely to lack of social skills development (Martinez-Pedraza & Carter, 2009).

There is approximately one fourth to one third of children with ASDs whose parents reported a significant loss or regression in language development. The regression characteristically occurs between 15-24 months of age (Lord et al., 2004; Luyster et al., 2005). Although, some parents reported normal development prior to regression, studies showed that some children with ASDs have subtle language and social impairments before the onset of regression (Richler et al., 2006; Werner & Dawson, 2005).

3.3 Restrictive interests, stereotypic and repetitive patterns of behaviours

Stereotypies and repetitive behaviours are not specific to children with ASDs. Children who have globally developmental delay (GDD) and children with sensory impairment may demonstrate stereotypies. Even in typically developing children, stereotypies may present e.g. flapping their hands when excited (Johnson, 2008). Stereotypies and repetitive behaviours in children with ASDs usually are not common in very young children (Charman & Baird, 2002; Cox et al., 1999; Moore & Goodson, 2003). Children with ASDs are preoccupied with sameness and routines, so interruption or changes in routine lead to tantrum and emotional disturbance. Some display sensory abnormalities: hypo- or hyper-responsive to sensory stimuli. Some children show an unusual and preoccupation with a topic of interest such as train schedules, solar system, dinosaurs, etc. However, this strong interest may not present in young children with ASDs. These patterns of behaviours vary among young individuals with ASDs. Therefore, diagnosis of ASDs in very young children should focus on social skills and language skills deficits rather than stereotypies and repetitive behaviours.

4. Screening tools for ASDs

The American Academy of Pediatrics (AAP) recommends ASDs screening in children age 18 and 24 months as part of developmental surveillance during regular health visits (Johnson & Myers, 2007). There are many valuable screening tools designed, such as the Checklist for Autism in Toddlers (CHAT) (Baron-Cohen et al., 1992; Baron-Cohen et al., 1996), the Modified Checklist for Autism in Toddlers (M-CHAT) (Kleinman et al., 2008; Robins et al., 2001), the Screening Test for Autism in Two-Year-Olds (STAT) (Stone et al., 2000) and the Pervasive Developmental Disorders Screening Test-II (PDDST-II) (Siegel, 2004). All of these tools, except the STAT, are designed as first-level screens (i.e. the tools are administered to all children to differentiate children who are at risk of ASDs from the general population).

Baron-Cohen *et al* conducted a study using the CHAT to administer in a primary health care setting to identify 18-month-old children at risk of ASDs. The study included both direct observation and a questionnaire for parents. The CHAT focuses on 3 key items which are gaze monitoring, protodeclarative pointing and pretend play. Findings from the study in the general population demonstrated that, the CHAT had a specificity of 98%-100% and a sensitivity of 18%-38% (Baird et al., 2000; Baron-Cohen et al., 1992; Baron-Cohen et al., 1996; Scambler et al., 2001). Attempts to improve sensitivity by modifying the cut-off criteria resulted in decrease in positive predictive value (from 75% to 5%). Overall, use of the CHAT as a screening tool remains problematic owing to low sensitivity (Bryson et al., 2003).

The M-CHAT is a screening tool for children 16 to 48 months and was developed to improve prediction of the CHAT. In the M-CHAT, there is no observation component, but includes a wider range of signs and symptoms of ASDs. This parental questionnaire consists of 23 (yes-no) items. Children who fail any three items or two critical items are considered to be at risk for ASDs. Items that were found to be the best predictors for ASDs were protodeclarative pointing, response to name, interest in peers, bringing things to show parents, following a point, and imitation. The reported sensitivity and specificity of the M-CHAT were around 89% and 93%, respectively (Dumont-Mathieu & Fein, 2005). However, the positive predictive value (PPV) was low (0.11±0.05) when it was used alone as a screen for ASDs in a community-based sample. The follow-up interview was reported to be able to significantly increase the PPV (Kleinman et al., 2008). Overall, the M-CHAT showed higher sensitivity than the CHAT and is possibly useful in identifying children in need of further assessments, but should not be used as a screen to exclude the possibility of ASDs (Eaves et al., 2006; Barbaro & Dissanayake, 2009).

The STAT is a second-level screen (that is, the tool is used to differentiate children who are at risk of ASDs from those at risk of other developmental disorders). It was designed to be used in children aged 2-3 years. The STAT includes 12 pass/fail items and is administered in a play-like setting in order to observe social-communicative behaviours. The test lasts approximately 20 minutes to complete. The estimated sensitivity and specificity were 95% and 73%, respectively (Stone et al., 2008). However, increased validity in larger studies and community-based samples are required.

The PDDST-II has both a first and second level screen versions. It is a parental questionnaire that can be used with children under 6 years of age. To date, the clinical validity remains unclear because it has not yet been published in a peer-reviewed journal (Volkmar et al., 2005).

5. Diagnostic instrument for ASDs

Currently, there are standardized instruments to facilitate diagnosis in ASDs. The Autism Diagnostic Interview - Revised (ADI-R) (Le Couteur et al., 2003; Lord et al., 1994) and the Autism Diagnostic Observation Schedule (ADOS) (Lord et al., 2000a) are well validated and currently their combination with clinical judgment based on the DSM-IV-TR criteria are considered as the "gold standard" for diagnosis of ASDs (Battaglia, 2007). However, these instruments should be used with caution in very young children or children with a mental age less than 24 months (Stone et al., 1999).

The ADOS is the most widely used standardized semistructured assessment of communication, social interaction and play. The scenarios for interaction with the child are used in the ADOS and require a well-trained interviewer. The ADOS consists of 4 modules devised for individuals with varying developmental and language level. Each module lasts approximately 40 minutes. The ADOS provides an algorithm to differentiate between autism, ASD and not ASD. Alpha coefficients are 0.86-0.91 for the social domain (across modules), 0.74-0.84 for communication, and 0.63-0.65 for repetitive behaviours (modules 1 and 2) (Lord et al., 2000a). In younger children, especially younger than 15 months of age, the sensitivity is excellent, the specificity is doubtful (Chawarska et al., 2007; Lord et al., 2000b; Risi et al., 2006). Luyster *et al* developed the toddler version of the ADOS (ADOS-Toddler Module or ADOS-T) which can be used for children under 30 months of age who have non-verbal mental ages of at least 12 months. The ADOS-T has acceptable internal consistency and excellent inter-rater and test-retest reliability (Luyster et al., 2009). However, larger samples of children and long follow-up studies need further replication.

The ADI-R is a standardized parental interview conducted by a trained interviewer. The interview covers the past developmental history and current functioning of individuals. The tool consists of 111 questions and takes about 2-3 hours. The ADI-R is designed to use in children about 4-5 years old. The ADI-R provides an algorithm to differentiate between autism and not autism. The ADI-R is reliable and valid. The inter-rater reliability on individual algorithm items ranges from 0.63 to 0.89.The internal consistency (alpha coefficients) is 0.69-0.95 (Lord et al., 1994). However, the time needed for administration precludes its use in clinical settings. Moreover, further study is needed for identifying ASDs in preschool children (Le Couteur et al., 2008; Mazefsky & Oswald, 2006; Risi et al., 2006).

The Developmental, Dimensional and Diagnostic Interview (3Di) is a new structured computerized interview for the diagnosis of ASDs and extends to co-morbid disorders. There are total 266 questions on autistic spectrum disorders (ASD) symptoms and 53 questions for an abbreviated interview. The questions in the interview are clustered according to domains of function: reciprocal social interaction skills, social expressiveness, use of language and other social communication skills, use of gesture and non-verbal play, and repetitive/stereotyped behaviours and routines. To reduce a risk of respondent bias, breaking down complex questions and scattering their components throughout the interview were done. A study reported that test-retest and inter-rater reliabilities were excellent. The sensitivity and specificity were estimated about 100% and 97%, respectively. Both the original 3di and the short version demonstrated high agreement with the ADI-R (Santosh et al., 2009; Skuse et al., 2004). Moreover, the short version takes less time to perform compared with the ADI-R. However, the study was limited to mild cases of ASDs; and so far limited numbers of young children have been tested.

The Autism Observation Scale for Infants (AOSI) (Bryson et al., 2008) is a diagnostic instrument that was developed for infants aged 6-18 months. The instrument consists of 18-item direct observational measure. Various activities were developed to assess the infant's target behaviours. These target behaviours are visual tracking and attentional disengagement; coordination of eye gaze and action; imitation; early social-affective and communicative behaviours; behavioural reactivity; and various sensory-motor behaviours. The inter-rater reliability ranges from 0.68 to 0.94 at 6, 12 and 18 months. Test-retest reliability is acceptable. The AOSI takes approximately 20 minutes to administer. Although, the AOSI is a useful diagnostic instrument for young children, it is not yet proposed to be used.

In brief, although there have been a number of screening and diagnostic instruments to facilitate ASDs diagnosis, a comprehensive evaluation for suspected ASDs should be performed. Such evaluations include a developmental history, parental interview, thorough physical examinations, clinical observations, developmental evaluations, assessment of the strengths and weaknesses of the child, assessment of family functioning, administration of standardized diagnostic instruments that operationalize the DSM criteria, and measures of cognitive and adaptive functions. Such comprehensive approaches together with early detection can lead to early intervention and result in improvement of the long-term functioning of children with ASDs.

6. Summary

Early detection of ASDs provides the best opportunity for early intervention, which results in significantly improved outcomes for children with ASDs. Awareness of the importance of early diagnosis and treatment has increased attention on knowledge of the very early manifestations of ASDs. Early manifestations include abnormalities in social interaction, communication and behaviours. Firstly, regarding social interaction, a lack of eye contact, orienting to name call, imitation, joint attention and limited responding to reciprocal play skills are the markers that should be of concern. Secondly, in the area of communication, any lack or delay of communication skills including verbal and non-verbal communication are indicative signs of ASDs. Lastly, the abnormal or unusual behaviours (i.e. repetitive and stereotypic behaviours, restrictive interests, preoccupied with sameness/ routine and sensory abnormalities) can be apparent in young children, however, these behaviours may not serve as important predictors of ASDs as the social and communication impairments.

Although, there are screening instruments to help identify children with ASDs in community-based samples, there is no screening instrument that provides adequate sensitivity and specificity for universal screening (Barbaro & Dissanayake, 2009). According to standardized diagnostic instruments, there have been many studies showing that the ADI-R and the ADOS have been well validated and are the instruments to accurately diagnose ASDs as early as 2 years. The combination of the ADOS and the ADI-R in conjunction with clinical diagnosis based on the DSM-IV-TR are recommended when diagnosing very young children with ASDs. In clinical practice where diagnostic instruments are not applicable, developmental surveillance with proper guidance is a recommended approach. Further prospective studies in young children should be conducted to provide evidence-based diagnosis for young children, especially under the age of two. Those developing research offer hope for better outcomes for children with ASDs.

7. Acknowledgments

We are very grateful to Dr. Suebwong Chuthapisith and Dr. Unchalee Lodin who proofread this article.

8. References

American Psychiatric Association. (2000). *Diagnostic and Statistical Manual of Mental Disorders, Fourth Edition, Text Revision (DSM-IV-TR)*. Washington DC: American Psychiatric Publishing, Inc.

Baird, G.; Charman, T.; Baron-Cohen, S.; Cox, A.; Swettenham, J.; Wheelwright, S. & Drew, A. (2000). A screening instrument for autism at 18 months of age: a 6-year follow-up study. *J Am. Acad. Child Adolesc. Psychiatry* 39(6):694-702

Baranek, GT. (1999). Autism during infancy: a retrospective video analysis of sensory-motor and social behaviors at 9-12 months of age. *J Autism Dev. Disord.* 29(3):213-24

Barbaro, J. & Dissanayake, C. (2009). Autism spectrum disorders in infancy and toddlerhood: a review of the evidence on early signs, early identification tools, and early diagnosis. *J Dev. Behav. Pediatr.* 30(5):447-59

Baron-Cohen, S.; Allen, J. & Gillberg, C. (1992). Can autism be detected at 18 months? The needle, the haystack, and the CHAT. *Br. J Psychiatry* 161:839-43

Baron-Cohen, S.; Cox, A.; Baird, G.; Swettenham, J.; Nightingale, N.; Morgan, K.; Drew, A. & Charman, T. (1996). Psychological markers in the detection of autism in infancy in a large population. *Br. J Psychiatry* 168(2):158-63

Battaglia, A. (2007). On the selection of patients with developmental delay/mental retardation and autism spectrum disorders for genetic studies. *Am. J Med Genet. A* 143A(8):789-90

Bryson, SE.; Zwaigenbaum, L.; McDermott, C.; Rombough, V. & Brian, J. (2008). The Autism Observation Scale for Infants: Scale development and reliability data. *J Autism Dev. Disord.* 38: 731-38

Bryson, SE.; Rogers, SJ. & Fombonne, E. (2003). Autism spectrum disorders: early detection, intervention, education, and psychopharmacological management. *Can. J Psychiatry* 48(8):506-16

Charman, T. (2003). Why is joint attention a pivotal skill in autism? *Philos. Trans. R. Soc. Lond B Biol. Sci.* 358(1430):315-24

Charman, T. & Baird, G. (2002). Practitioner review: Diagnosis of autism spectrum disorder in 2- and 3-year-old children. *J Child Psychol. Psychiatry* 43(3):289-305

Chawarska, K.; Klin, A.; Paul, R. & Volkmar, F. (2007). Autism spectrum disorder in the second year: stability and change in syndrome expression. *J Child Psychol. Psychiatry* 48(2):128-38

Cox, A.; Klein, K.; Charman, T.; Baird, G.; Baron-Cohen, S.; Swettenham, J.; Drew, A. & Wheelwright, S. (1999). Autism spectrum disorders at 20 and 42 months of age: stability of clinical and ADI-R diagnosis. *J Child Psychol. Psychiatry* 40(5):719-32

Dawson, G.; Munson, J.; Estes, A.; Osterling, J.; McPartland, J.; Toth, K.; Carver, L. & Abbott, R. (2002). Neurocognitive function and joint attention ability in young children with autism spectrum disorder versus developmental delay. *Child Dev.* 73(2):345-58

De Giacomo A, Fombonne E. 1998. Parental recognition of developmental abnormalities in autism. *Eur. Child Adolesc. Psychiatry* 7(3):131-6

Dumont-Mathieu, T. & Fein, D. (2005). Screening for autism in young children: The Modified Checklist for Autism in Toddlers (M-CHAT) and other measures. *Ment. Retard. Dev. Disabil. Res. Rev.* 11(3):253-62

Eaves, LC.; Wingert, H. & Ho, HH. (2006). Screening for autism: agreement with diagnosis. *Autism* 10(3):229-42

Gray, KM. & Tonge, BJ. (2001). Are there early features of autism in infants and preschool children? *J Paediatr. Child Health* 37(3):221-6

Johnson, CP. (2008). Recognition of autism before age 2 years. *Pediatr. Rev.* 29(3):86-96

Johnson, CP. & Myers, SM. (2007). Identification and evaluation of children with autism spectrum disorders. Pediatrics 120(5):1183-215

Kleinman, JM.; Robins, DL.; Ventola, PE.; Pandey, J.; Boorstein, HC.; Esser, EL.; Wilson, LB.; Rosenthal, MA.; Sutera, S.; Verbalis, AD.; Barton, M.; Hodgson, S.; Green, J.; Dumont-Mathieu, T.; Volkmar, F.; Chawarska, K.; Klin, A. & Fein, D. (2008). The modified checklist for autism in toddlers: a follow-up study investigating the early detection of autism spectrum disorders. *J Autism Dev. Disord.* 38(5):827-39

Le Couteur, A.; Haden, G.; Hammal, D. & McConachie, H. (2008). Diagnosing autism spectrum disorders in pre-school children using two standardised assessment instruments: the ADI-R and the ADOS. *J Autism Dev. Disord.* 38(2):362-72

Le Couteur, A.; Lord, C. & Rutter, M. (2003). *The Autism Diagnostic Interview-Revised (ADI-R).* Los Angeles, CA: Western Psychological Services.

Lord, C.; Risi, S.; Lambrecht, L.; Cook, EH Jr.; Leventhal, BL.; DiLavore, PC.; Pickles, A. & Rutter, M. (2000a). The autism diagnostic observation schedule-generic: a standard measure of social and communication deficits associated with the spectrum of autism. *J. Autism Dev. Disord.* 30(3):205-23

Lord, C.; Risi, S. & Wetherby, AM., et al. (2000b). Diagnosis of autism spectrum disorders in young children. In *Autism spectrum disorders: a transactional development perspective,* ed. Wetherby AM, Prizant BM,11-30 pp. Baltimore: Paul H Brookes Publishing. 11-30 pp.

Lord, C.; Rutter, M. & Le Couteur, A. (1994). Autism Diagnostic Interview-Revised: a revised version of a diagnostic interview for caregivers of individuals with possible pervasive developmental disorders. *J Autism Dev. Disord.* 24(5):659-85

Lord, C.; Shulman, C. & DiLavore, P. (2004). Regression and word loss in autistic spectrum disorders. *J Child Psychol. Psychiatry* 45(5):936-55

Luyster, R.; Gotham, K.; Guthrie, W.; Coffing, M.; Petrak, R.; Pierce, K.; Bishop, S.; Esler, A.; Hus, V.; Oti, R.; Richler, J.; Risi, S. & Lord, C. (2009). The Autism Diagnostic

Observation Schedule-toddler module: a new module of a standardized diagnostic measure for autism spectrum disorders. *J Autism Dev. Disord.* 39(9):1305-20

Luyster, R.; Richler, J.; Risi, S.; Hsu, WL.; Dawson, G.; Bernier, R.; Dunn, M.; Hepburn, S.; Hyman, SL.; McMahon, WM.; Goudie-Nice, J.; Minshew, N.; Rogers, S.; Sigman, M.; Spence, MA.; Goldberg, WA.; Tager-Flusberg, H.; Volkmar, FR. & Lord, C. (2005). Early regression in social communication in autism spectrum disorders: a CPEA Study. *Dev. Neuropsychol.* 27(3):311-36

Martinez-Pedraza, FL. & Carter, AS. (2009). Autism spectrum disorders in young children. *Child Adolesc. Psychiatr Clin. N. Am.* 18(3):645-63

Mazefsky, CA. & Oswald, DP. (2006). The discriminative ability and diagnostic utility of the ADOS-G, ADI-R, and GARS for children in a clinical setting. *Autism* 10(6):533-49

McDonough, L.; Stahmer, A.; Schreibman, L. & Thompson, SJ. (1997). Deficits, delays, and distractions: an evaluation of symbolic play and memory in children with autism. *Dev. Psychopathol.* 9(1):17-41

Moore, V. & Goodson, S. (2003). How well does early diagnosis of autism stand the test of time? Follow-up study of children assessed for autism at age 2 and development of an early diagnostic service. *Autism* 7(1):47-63

Osterling, J. & Dawson, G. (1994). Early recognition of children with autism: a study of first birthday home videotapes. *J Autism Dev. Disord.* 24(3):247-57

Rapin, I. & Tuchman, RF. (2008). Autism: definition, neurobiology, screening, diagnosis. *Pediatr. Clin. North Am.* 55(5):1129-46, viii

Richler, J.; Luyster, R.; Risi, S.; Hsu, WL.; Dawson, G.; Bernier, R.; Dunn, M.; Hepburn, S.; Hyman, SL.; McMahon, WM.; Goudie-Nice, J.; Minshew, N.; Rogers, S.; Sigman, M.; Spence, MA.; Goldberg, WA.; Tager-Flusberg, H.; Volkmar, FR. & Lord, C. (2006). Is there a 'regressive phenotype' of Autism Spectrum Disorder associated with the measles-mumps-rubella vaccine? A CPEA Study. *J Autism Dev. Disord.* 36(3):299-316

Risi, S.; Lord, C.; Gotham, K.; Corsello, C.; Chrysler, C.; Szatmari, P.; Cook, EH Jr.; Leventhal, BL. & Pickles, A. (2006). Combining information from multiple sources in the diagnosis of autism spectrum disorders. *J Am. Acad. Child Adolesc. Psychiatry* 45(9):1094-103

Robins, DL.; Fein, D.; Barton, ML. & Green, JA. (2001). The Modified Checklist for Autism in Toddlers: an initial study investigating the early detection of autism and pervasive developmental disorders. *J Autism Dev. Disord.* 31(2):131-44

Rutter, M. (2006). Autism: its recognition, early diagnosis, and service implications. *J Dev. Behav. Pediatr.* 27(2 Suppl):S54-S58

Rutter, M.; Moffitt, TE. & Caspi, A. (2006). Gene-environment interplay and psychopathology: multiple varieties but real effects. *J Child Psychol. Psychiatry* 47(3-4):226-61

Santosh, PJ.; Mandy, WP.; Puura, K.; Kaartinen, M.; Warrington, R. & Skuse, DH. (2009). The construction and validation of a short form of the developmental, diagnostic and dimensional interview. *Eur. Child Adolesc. Psychiatry* 18(8):521-4

Scambler, D.; Rogers, SJ. & Wehner, EA. (2001). Can the checklist for autism in toddlers differentiate young children with autism from those with developmental delays? *J Am. Acad. Child Adolesc. Psychiatry* 40(12):1457-63

Siegel, B. (2004). *Pervasive developmental disorders screening test-ii (pddst-ii)*. San Antonio: Harcourt.

Sigman, M.; Ruskin, E.; Arbeile, S.; Corona, R.; Dissanayake, C.; Espinosa, M.; Kim, N.; Lopez, A. & Zierhut, C. (1999). Continuity and change in the social competence of children with autism, Down syndrome, and developmental delays. *Monogr Soc. Res. Child Dev.* 64(1):1-114

Skuse, D.; Warrington, R.; Bishop, D.; Chowdhury, U.; Lau, J.; Mandy, W. & Place, M. (2004). The developmental, dimensional and diagnostic interview (3di): a novel computerized assessment for autism spectrum disorders. *J Am. Acad. Child Adolesc. Psychiatry* 43(5):548-58

Stone, WL.; Coonrod, EE. & Ousley, OY. (2000). Brief report: screening tool for autism in two-year-olds (STAT): development and preliminary data. *J Autism Dev. Disord.* 30(6):607-12

Stone, WL.; Lee, EB.; Ashford, L.; Brissie, J.; Hepburn, SL.; Coonrod, EE. & Weiss, BH. (1999). Can autism be diagnosed accurately in children under 3 years? *J Child Psychol. Psychiatry* 40(2):219-26

Stone, WL.; Lemanek, KL.; Fishel, PT.; Fernandez, MC. & Altemeier, WA. (1990). Play and imitation skills in the diagnosis of autism in young children. *Pediatrics* 86(2):267-72

Stone, WL.; McMahon, CR. & Henderson, LM. (2008). Use of the Screening Tool for Autism in Two-Year-Olds (STAT) for children under 24 months: an exploratory study. *Autism* 12(5):557-73

Szatmari, P. (1999). Heterogeneity and the genetics of autism. *J. Psychiatry Neurosci.* 24(2):159-65

Turner, LM.; Stone, WL.; Pozdol, SL. & Coonrod, EE. (2006). Follow-up of children with autism spectrum disorders from age 2 to age 9. *Autism* 10(3):243-65

Turner, M. (1999). Annotation: Repetitive behaviour in autism: a review of psychological research. *J Child Psychol. Psychiatry* 40(6):839-49

Volkmar, F.; Chawarska, K. & Klin, A. (2005). Autism in infancy and early childhood. *Annu. Rev. Psychol.* 56:315-36

Werner, E. & Dawson, G. (2005). Validation of the phenomenon of autistic regression using home videotapes. *Arch. Gen. Psychiatry* 62(8):889-95

Wetherby, AM.; Prizant, BM. & Schuler, AL. (2000). Understanding the nature of communication and language impairments. In *Autistic Spectrum Disorders: A developmental transactional perspective*, ed. Wetherby AM, Prizant BM,109-141 pp. Baltimore: Paul H. Brookes. 109-141 pp.

World Health Organization (WHO). (1992). *The ICD-10 Classification of Mental and Behavioural Disorders*. Geneva: WHO.

Young, RL.; Brewer, N. & Pattison, C. (2003). Parental identification of early behavioural abnormalities in children with autistic disorder. *Autism* 7(2):125-43

Zwaigenbaum, L.; Bryson, S.; Rogers, T.; Roberts, W.; Brian, J. & Szatmari, P. (2005). Behavioral manifestations of autism in the first year of life. *Int. J Dev. Neurosci.* 23(2-3):143-52

Part 2

Nosology and Diagnostic Criteria: What Makes Sense and Can Genetics Help?

2

Pervasive Developmental Disorder- not Otherwise Specified: Specifying and Differentiating

Koray Karabekiroglu
Ondokuz Mayis University
Turkey

1. Introduction

Pervasive Developmental Disorders (PDD), also called Autism Spectrum Disorders (ASD), are defined in terms of abnormalities in social and communication development in the presence of marked repetitive behaviour and narrow interests (APA, 1994). The *DSM-IV* (APA, 1994) and ICD-10 (WHO, 1993) provide diagnostic criteria for autism and related disorders such as Asperger syndrome (AS), Rett's, and childhood disintegrative disorder. Unfortunately, the diagnostic category of pervasive development disorder-not otherwise specified (PDD-NOS) does not have specific criteria and is often seen as a catchall diagnosis for children who do not fit the criteria for one of the other pervasive developmental disorders (Filipek et al., 1999).

According to Cohen & Volkmar (2005) classification systems should aim at improving communication, through their features (internal consistency, use easiness, good definition of categories) and being widely accepted. The accuracy of early diagnosis, as well as developmental pathways that are observed in young children with ASD have both theoretical and practical importance (Luyster et al., 2005). An empirically developed dimensional approach that defines the spectrum on multiple dimensions may offer several advantages. It may, for example, result in more correspondence between the results of genetic research and the phenotype of autistic disorders, provided the pathology can be summarized by empirical and valid behavior dimensions (Volkmar et al., 2004; van Lang et al., 2006; Hus et al., 2007).

It is now well recognized that children with PDD vary in the number and severity of symptoms (Szatmari et al., 2002). In DSM-IV, a diagnostic category within PDD, which is called "pervasive developmental disorder-not otherwise specified" (PDD-NOS), defines children with symptoms such as restricted social interaction, poor verbal and non-verbal communication skills, strict and/or stereotypical behaviors but without full diagnostic criteria of autism (APA, 1994). Epidemiological data suggest that PDD-NOS is at least twice as common as autism in the general community (Chakrabatri & Fombonne, 2001). One or more of the following conditions may lead to a PDD-NOS diagnosis (1) onset of the disorder after 3 years of age, (2) atypical symptoms with regard to the 12 criteria of autism specified in DSM-IV, (3) fewer than 6 criteria and thus subtreshold (Walker et al., 2004). A categorical system like DSM-IV can be very useful for diagnosing prototypic manifestations of a disorder, but it is less useful in encompassing what may be, in its broader manifestations, a "spectrum disorder" (Tanguay, 2004). An assumption of the autism-spectrum model is that autism conditions lie on a continuum of social-communication skills (Baron-Cohen et al., 2001; Wakabayashi et al., 2007). A continuum view shifts us away from categorical diagnosis and towards a quantitative approach.

Diagnostic agreement for PDD-NOS is generally considered to be weak (Tanguay, 2004). Walker and colleagues presented compelling evidence, both from the literature and from their study, that attempting to improve the DSM-IV criteria for PDD-NOS can be quite frustrating (Walker et al., 2004). Many of the symptoms of PDD-NOS can occur in non-PDD conditions, such as severe mental retardation or language delay, and they may present with similar developmental history (Bishop et al., 2006). Furthermore, clinical presentation of PDD-NOS may resemble presenting symptoms in high functioning autism, Asperger's disorder, reactive attachment disorder, and psychotic disorders, and the differential diagnosis may be highly complicated.

Studies on the distinction between Autistic Disorder (AD) and Pervasive Developmental Disorder Not Otherwise Specified (PDD-NOS) have been inconclusive (Snow & Lecavalier, 2011). The field is in need of more studies examining subtype differences. As the diagnostic validity of PDD-NOS is still open to question, and to explore proposed underlying factors, we have to assign cases based on a valid clinical assessment. Therefore, we still need to investigate further the clinical features of children with PDD-NOS that distinguish them from children with autism and other non-PDD conditions.

2. Autism, PDD-NOS, and ADHD

Barkley (1990) reported that it is common for children with PDD-NOS to be initially given a diagnosis of Attention Deficit/Hyperactivity Disorder (ADHD). Jensen et al., (1997) reported that 74% of the children in their study diagnosed with PDD-NOS were originally diagnosed with ADHD. Another study showed that children with PDD-NOS and ADHD did not differ from each other with respect to total number of autistic symptoms, general psychopathology, or attention difficulties (Luteijn et al., 2000). Methods for differentiating PDD-NOS from the non-PDD disorders, such as attention deficit hyperactivity disorder (ADHD), are not well established. Several investigators concluded that it is difficult to make a distinction between ADHD and PDD by using the present diagnostic criteria in DSM-IV (Bryson et al., 2008; Gökler et al., 2004). The characteristics that differentiated children with PDD-NOS from those with autism and non-PDD disorders were also explored by Buitelaar et al., (1999). Four criteria discriminated autism from PDD-NOS most effectively: children with autism more often demonstrated restricted patterns of interest, lacked varied make-believe play, failed to use nonverbal behavior, and had an earlier age of onset. In another study (Allen et al., 2001), the PDD-NOS group (including both high- and low-functioning children) did not differ significantly from the autism or non-PDD groups on measures of language or adaptive functioning but did show less restricted stereotyped behaviors than the high-functioning autism group.

In a very recent study (Snow & Lecavalier, 2011), authors examined the validity of PDD NOS by comparing it to autistic disorder (AD) and other developmental disorders (DD) on parent-reported behavior problems. Fifty-four children with PDD-NOS were individually matched on age and nonverbal IQ to 54 children with AD and 54 children with DD. The only difference between PDD-NOS and AD groups was higher scores in the PDD-NOS group on two items measuring Anxiety/Depression. Cognitive functioning may be a more salient variable than subtype when studying psychopathology in individuals with ASDs.

In a study (Karabekiroglu & Akbas, in press) designed to explore whether PDD-NOS encompassed a distinct cluster of symptoms and clinical profile or not, we investigated differential features of PDD-NOS such as presenting symptoms, developmental history, and comorbidity with respect to autism and ADHD. The study involved 188 children (PDD-NOS n=94; ADHD n=47; autism n=47) (male n=150, female n=38) who were 5.5(±2.5) years old on average (range 2-11 yrs.). The children with Asperger Syndrome were excluded. Preliminary PDD-NOS screening scale (PPSSS) was developed based on the 'presenting' symptoms of PDD-NOS that were systematically collected in a pilot group of children (Table 1).

The clinical diagnoses and comorbidities were based on the comprehensive mental status examination, Schedule for Affective Disorders and Schizophrenia for School Age Children-Present and Lifetime Version-Turkish Version (K-SADS-PL-T), and the consensus between two child and adolescent psychiatry specialists. The prevelance rates of the most common presenting symptoms in the PDD-NOS and autism groups showed a similar pattern of distribution from most common to the least (Figure 1), even when the results were corrected for age. However, almost all of these symptoms are reported significantly less in prevalence in the PDD-NOS group.

In this study, ADHD was also explored as a co-morbid diagnosis; 38.3% of the children in the PDD-NOS group and 53.2% of the children with autism fullfilled ADHD criteria (p>.05). Compared with children in the PDD-NOS group, children in the ADHD group had significantly higher rates of co-morbid disruptive behavior disorders (27.6% vs. 9.6%), learning disorders (14.9% vs. 5.3%), elimination disorders (12.8% vs. 2.1%), tic disorders (8.5% vs. 2.1%), social anxiety disorder (8.5% vs. 2.1%) and lower rates of co-morbid obsessive compulsive disorder (2.1% vs. 23.4%). The rates of other co-morbid disorders, such as depression, language disorders, and sleep disorders, were found to be similar across diagnostic groups. The findings of this study reveal that the PDD-NOS group had a high number of features in common with the autism and the ADHD groups, in terms of presenting and/or reported symptoms and developmental history. Similar to previous studies (Volkmar, et al 1993), gender distribution was similar for all groups (in each group more than 75% of the patients were male). A recent study has suggested that approximately 70% of children with ASDs have at least one comorbid psychiatric disorder (Simonoff et al., 2008). The most prevalent comorbid disorders were anxiety disorders (42%), oppositional or conduct disorders (30%), and ADHD (28%).

In our study (Karabekiroglu & Akbas, in press), as shown in Table 1 and Figure 1, the prevelance rates of the most common presenting symptoms in the PDD-NOS and autism groups had a similar pattern of distribution from more to less common. However, almost all of these symptoms were reported significantly less in children diagnosed with PDD-NOS than children with autism. The autism and the PDD-NOS shared a common clinical symptom profile on the first clinical admission. On the other hand, the children with ADHD had a distinct set of symptoms. The results suggest that PDD-NOS may be assumed as a quantitative partial subtype of autism, and it represents a less severe form that lies on a continuum of social-communication skills.

Preliminary PDD-NOS Symptom Screening Scale (PPSSS) Items	Presence of the symptoms (percentages)				
	PDD-NOS (1)	**Autism (2)**	**ADHD (3)**	Overall significance (p value)	Source of significance
1. poor social interaction	59.6	97.9	8.5	**<.001**	1:2; 1:3; 2:3
2. hyperactivity	56.4	80.9	89.4	**<.001**	1:2; 1:3
3. not speaking/ language retardation	53.2	97.9	6.4	**<.001**	1:2; 1:3; 2:3
4. aggressiveness	33.0	46.8	61.7	N.S.	
5. stubbornness	31.9	46.8	44.7	N.S.	
6. inattentiveness	30.9	66.0	91.5	**<.001**	1:2; 1:3; 2:3
7. obsessions	29.8	27.7	14.9	N.S.	
8. not responsive to social stimuli	25.5	95.7	40.4	**<.001**	1:2; 2:3
9. stereotypies	24.5	59.6	6.4	**<.001**	1:2; 1:3; 2:3
10. impatience and/or impulsiveness	23.4	48.9	78.7	**<.001**	1:2; 1:3; 2:3
11. fastidiousness, choosyness	23.4	10.6	10.6	N.S.	
12. echolalia	22.3	14.9	-	N.S.	
13. highly interested in television	20.2	46.8	17.0	**<.001**	1:2; 2:3
14. conduct problems	21.3	36.2	40.4	N.S.	
15. articulation and/or prosody problems	18.1	8.5	4.3	N.S.	
16. lack of eye contact	14.9	59.6	-	**<.001**	1:2; 1:3; 2:3
17. multiple fears	14.9	8.5	17.0	N.S.	
18. sleep problems	14.9	10.6	27.7	N.S.	
19. tactile oversensitivity	12.8	25.5	6.4	N.S.	
20. confusing pronouns	11.7	12.8	-	N.S.	
21. shyness	11.7	6.4	17.0	N.S.	
22. emotional lability	11.7	-	27.7	**<.001**	1:3; 2:3
23. tics	10.6	4.3	10.6	N.S.	
24. poor appetite	7.4	25.5	21.3	N.S.	
25. inappropriate laughing	4.3	17.0	2.1	N.S.	
26. persistence with sameness	2.1	12.8	6.4	N.S.	
27. frequent startles	1.1	10.6	8.5	N.S.	

Table 1. Preliminary PDD-NOS Symptom Screening Scale (PPSSS) item distributions of patients in each diagnosis group

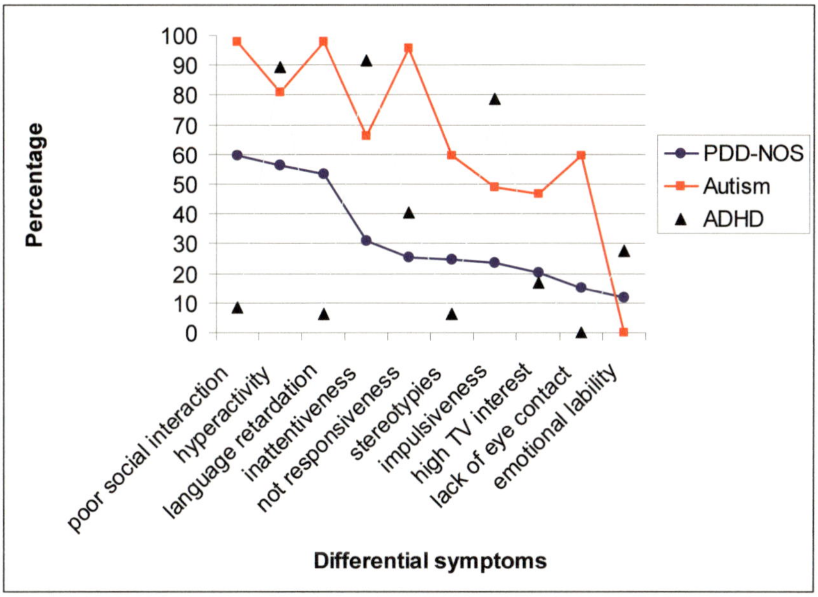

Fig. 1. The significantly discriminative symptom percentages of the diagnostic groups

3. Cluster and factor analysis

To identify ASD subgroups, several investigators used cluster and factor analysis based on social functioning, intelligence, developmental milestones, and so forth. Various clusters were reported (Eaves et al., 1994; Prior et al., 1998; Sevin et al., 1995; Waterhouse et al., 1996; Wing & Gould, 1979). But these findings were not replicated and the clusters identified were not adopted or replicated in later studies. Despite several studies with ASD, clinical validity and differential features of PDD-NOS are yet to be consistently established. A very recent study (Shumway et al., 2011) examined the relationship between onset status and current functioning using a recently proposed onset classification system in 272 young children with autism spectrum disorder (ASD). Participants were classified into one of the following groups, based on parent report using the Autism Diagnostic Interview—Revised: Early Onset (symptoms by 12 months, no loss), Delay and Regression (symptoms by 12 months plus loss), Plateau (no early symptoms or loss), and Regression (no early symptoms, followed by loss). Findings indicate that current functioning does not differ according to onset pattern, calling into question the use of onset categorizations for prognostic purposes in children with ASD.

A previous study performed a factor analysis on a sample of variant categories of PDD, and two factors emerged. One factor represented autistic symptoms and another represented level of functioning (Szatmari et al., 2002). More recent studies used a factor analytic approach based on particular diagnostic instruments, such as the Autism Diagnostic Interview-Revised (ADI-R) and the Autism Diagnostic Observation Schedule (ADOS) (Tadevosyan-Layfer et al., 2003; Tanguay, 2004). The results suggested that there is a developmental continuum from affective reciprocity to emotional joint attention to verbal joint attention and to intuitive social knowledge (Tanguay, 2004). Tadevosyan-Layfer et al. (2003) found six factors: spoken language, compulsions, developmental milestones, savant skills, sensory aversion, and social intent.

In our study (Karabekiroglu & Akbas, in press), including all subjects in all diagnostic groups (PDD-NOS n=94; ADHD n=47; autism n=47) (male n=150, female n=38) who were 5.5(±2.5) years old on average (range 2-11 yrs.), a principal axis factor analysis with Promax rotation revealed ten factors; seven were found to be discriminative (Table 2, Figure 2). We

Item	Factor Loading										Communality
	1	2	3	4	5	6	7	8	9	10	
1. lack of eye contact	**.69**										.43
2. stereotypies	**.57**										.37
3. inappropriate laughing	**.52**										.26
4. frequent startles	**.46**					.35					.41
5. highly interestedness in TV	**.41**						-.33				.29
6. tactile oversensitiveness	**.35**										.24
7. poor social interaction		**.85**									.48
8. language retardation		**.51**	.31								.42
9. not responsive		**.48**									.33
10. emotional lability		-.42									.26
11. inattentiveness			**.68**								.34
12. hyperactivity			**.53**								.29
13. impatiente, impulsiveness			**.48**	.30							.39
14. aggressiveness				**.76**							.30
15. conduct problems				**.58**							.32
16. confusing pronouns					**.75**						.27
17. echolalia			-.32		**.46**						.28
18. articulation/ prosody problems					**.39**						.17
19. several fears						**.66**					.27
20. shyness						**.53**					.22
21. fastidiousness, choosyness							**.54**				.20
22. obsessions							**.54**	.35			.35
23. poor appetite								**.53**			.28
24. stubbornness								**.51**			.29
25. persistence with sameness								**.30**			.28
26. sleep problems								.31	**.92**		.32
27. tics										**.74**	.20
Eigenvalues	3.28	2.88	1.94	1.65	1.37	1.36	1.31	1.41	1.11	1.05	
% of variance	12.15	10.67	7.18	6.12	5.07	5.03	4.84	4.22	4.10	3.90	

Note. Loadings <.30 are omitted. Adopted items into the factors are shown bold.

Table 2. PPSSS items and factor loadings for the rotated ten factors

retained all components with eigenvalue (a measure of explained variance) greater than unity. Ten factors had eigenvalues greater than 1.0, which is a common criterion for a factor to be useful. When ten factors were requested, Kaiser-Meyer-OIkin (KMO) measure was adequate (.66), and Bartlett's Test of Spherity was significant ($p<.001$). These measures mean that the variables are correlated highly enough to provide a reasonable basis for factor analysis. We considered all variables with factor loadings 0.3 or larger in the appropriate factor matrices to define the underlying factor and we took these variables as a cluster of variables for the factor. The two rotation procedures produced similar results. When there were differences, we took the Promax solution as the preferred one. After rotation, ten factors accounted for 66.3% of the variance.

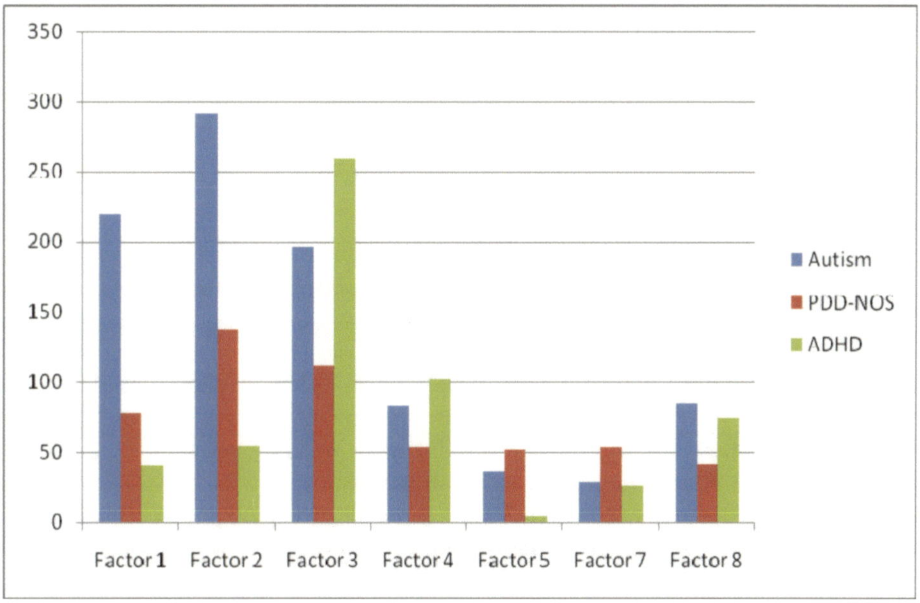

Factor 1 includes "lack of eye contact", "stereotypies", "inappropriate laughing", "frequent startles", "highly interestedness in TV", and "tactile oversensitiveness";
Factor 2 includes "poor social interaction", "language retardation", and "not responsive"
Factor 3 includes "inattentiveness", "hyperactivity", and "impatiente, impulsiveness"
Factor 4 includes "aggressiveness" and "conduct problems"
Factor 5 includes "confusing pronouns", "echolalia", and "articulation/ prosody problems"
Factor 7 includes "fastidiousness, choosiness" and "obsessions"
Factor 8 includes "poor appetite", "stubbornness", and "persistence with sameness"

Fig. 2. The significantly discriminative factors of the diagnostic groups.

We found significant differences in the toal number of symptoms between three diagnostic groups in the factors 1 ($p<.001$), 2 ($p<.001$), 3 ($p<.001$), 4 ($p=.004$), 5 ($p<.001$), 7 ($p=.026$), and 8 ($p=.006$). The scores in the factors 1, 2, 3, and 8 were significantly higher in the autism group compared to the PDD-NOS group. The scores in the factors 1, 2, 5, and 7 were significantly higher in the PDD-NOS group compared to the ADHD group. Inversely, the scores in the factors 3, 4, and 8 were significantly higher in the ADHD group compared to the PDD-NOS group (Figure 2).

Based on the assumption that the items were predicted to index three constructs: symptoms related to autism, ADHD, and PDD-NOS, in a further analysis three factors were requested (Karabekiroglu & Akbas, In press). The first factor seemed to index core autism spectrum, the second factor, disruptive behaviors spectrum, and the third factor seemed to index symptoms to be interpreted as anxiety spectrum. Four in twenty-seven items do not seem to load with any of the factors. When the total number of the symptoms in each factor were compared between the diagnostic groups, the *core autism spectrum* and the *disruptive behavior spectrum* factors revealed significant differences between the groups ($p<.001$). Post-hoc analysis showed that in the *core autism spectrum factor*, the autistic group had significantly more symptoms than the PDD-NOS group (4.87 vs. 2.14) ($p<.001$), and the PDD-NOS group had significantly more symptoms than the ADHD group (2.14 vs. 0.81) ($p<.001$). On the other hand, on the *disruptive behavior spectrum* factor, both the ADHD (3.62 vs. 2.19) ($p<.001$) and the autistic groups (4.55 vs. 2.19) ($p<.001$) had significantly more symptoms than the PDD-NOS group. The *anxiety spectrum* factor did not reveal a significant difference between diagnostic groups.

4. Discussion

Because the diagnostic agreement for PDD-NOS was generally considered to be weak (Tanguay 2004, Walker et al. 2004), and differentiation of PDD-NOS from the non-PDD disorders, such as ADHD was not well-defined, we conducted a factor analysis including the data from all three diagnosis groups (Autism, PDD-NOS, and ADHD) (Karabekiroglu & akbas, in press). A factor analysis revealed three symptom clusters, *core autistic spectrum, disruptive behavior spectrum,* and *anxiety spectrum.* As would be expected, the children with autism had higher rates of symptoms in the *autistic spectrum* factor and the children with ADHD had higher rates of symptoms in the *disruptive behavior spectrum* factor. The PDD-NOS group had lower rates of symptoms on both factors.

In a recent study (Kamp-Becker et al., 2009), the dimensional structure of higher functioning autism phenotype was investigated by factor analysis. The goal of this study was to identify the degree to which early symptoms of autism (measured using the ADI-R) could be predictive of the current symptoms of autism as identified using the ADOS, the adaptive behavior scales, IQ scores and theory of mind scores. The authors reported that the social interaction and communication domains were closely related to one factor namely: Social communication. An additional factor implies anxious and compulsive behavior which is associated with current social communication functioning. Another study compared the behavioral symptomatology in 26 children and adolescents with autism and 25 children and adolescents with PDD-NOS (Pearson et al., 2006). Relative to individuals with PDD-NOS, those with autism had more symptoms of depression, social withdrawal, atypical behavior, and immature social skills, and fewer family problems. These differences remained even when group differences in intellectual ability were controlled statistically. No group differences emerged in somatization, anxiety, or hyperactivity. Their findings suggested that, although both groups demonstrated considerable evidence of behavioral and emotional problems, those with autism were at particularly high risk for co-morbid behavioral and emotional disabilities (Pearson et al., 2006).

In a recent study (Mandy et al., 2011) authors aimed, first, to improve the reliability and replicability of PDD-NOS by operationalizing its DSM-IV-TR description and, second, to test its validity through comparison with autistic disorder (AD) and Asperger's disorder (AsD).

In a sample of 256 young people (mean age: 9.1 years) [AD (n:97), AsD (n:93) and PDD-NOS (n:66)], groups were compared on independent measures of core PDD symptomatology, associated autistic features, and intelligence. Contrary to the assumption that PDD-NOS is heterogeneous, almost all (97%) of those with PDD-NOS had one distinct symptom pattern, namely impairments in social reciprocity and communication, without significant repetitive and stereotyped behaviors (RSB). Compared to AD and AsD, they had comparably severe but more circumscribed social communication difficulties, with fewer non-social features of autism, such as sensory, feeding and visuo-spatial problems. These individuals appear to have a distinct variant of autism that does not merely sit at the less severe end of the same continuum of symptoms.

The symptoms of ASD may change with development (Luyster et al., 2005). PDD-NOS has been assumed significantly less stable as a diagnosis (Lord et al., 2006). In a study (Kleinman et al., 2008), 77 children received a diagnostic and developmental evaluation between 16 and 35 months and also between 42 and 82 months. Diagnoses based on clinical judgment, Childhood Autism Rating Scale, and the Autism Diagnostic Observation Schedule were stable over time. Diagnoses made using the Autism Diagnostic Interview were slightly less stable. According to clinical judgment, 15 children (19%) moved off the autism spectrum by the second evaluation; none moved onto the spectrum. Results indicate diagnostic stability at acceptable levels for diagnoses made at age 2. Nevertheless, diagnoses of autism and PDD-NOS by experienced clinicians on the basis of multiple measures were valid and reliable over time (Lord et al., 2006). If a child is given an ASD diagnosis (either autism or PDD-NOS) at age 2 years, it is highly likely to apply at age 9, although there may be some shifting within the range of ASD diagnostic categories (Lord et al., 2006). Generally, it appears that the overall picture of development for autism and PDD-NOS is similar, with most children experiencing continued impairment. Based on these two studies, there does not appear to be evidence for qualitatively discrete groups (i.e., autism versus PDD-NOS), but differences appear to be quantitative (Lord et al., 2006; Turner, et al., 2006).

A recent meta-analysis (Rondeau et al., 2010) conducted on the eight longitudinal studies on PDD-NOS that have been published from 1996 to 2009 showed that PDD-NOS diagnosis was less stable than autistic disorder diagnosis. When established before 36 months, the overall stability rate was 35% at 3-year follow-up. Consistent with the previous literature on the reliability of the PDD-NOS diagnosis in young children, our metaanalysis did not support the discriminant and predictive validity of this category. Thus, from a clinical standpoint, children whose PDD-NOS diagnosis was established before 36 months should be re-assessed at a later age (Rondeau et al., 2010).

Similar to previous reports (Allen et al., 2001, deBruin et al., 2006, Matson, et al., 2007, Szatmari et al. 2002), in our study (Karabekiroglu & Akbas, in press) mental retardation was significantly more prevalent in the autism than in the PDD-NOS or ADHD groups. Several investigators suggested that exploring the presence of mental retardation may be more useful in terms of planning treatment and predicting outcome than a classification based on symptom number alone (Szatmari et al., 2002). However, IQ may be a poor measure of level of functioning, based as it is on performance in a highly artificial setting (Szatmari et al. 2002). In a study (Scheirs & Timmers, 2009) an attempt was made to distinguish among the three groups (ADHD, PDD-NOS, and ADHD plus PDD-NOS) on the basis of intelligence (WISC-III) profiles. It was found that the PDD-NOS group had higher verbal and performance IQ's, as well as higher WISC-III index scores than the ADHD group. Subtests

Block Design and Mazes discriminated best. It was concluded that based on intelligence scores, only PDD-NOS and ADHD emerged as distinct categories, whereas the combined diagnosis did not. Allen et al. (2001) compared 18 preschool children with PDD-NOS to 176 children with autistic disorder and 311 non-autistic children with developmental language disorders (DLD) (N = 201) or low IQ (N = 110). The children with PDD-NOS did not differ significantly from either the children with autism or the children with DLD in verbal and adaptive skills. They suggested that the similarity of PDD-NOS children to autistic children in maladaptive behaviors and an intermediate position between autistic and DLD groups on virtually all measures helped to explain the difficulty clinicians encounter in classifying children with PDD-NOS (Allen et al., 2001).

Rates of comorbid psychiatric conditions in children with PDD-NOS are hardly available, although these conditions are often considered as more responsive to treatment than the core symptoms of PDD-NOS (deBruin et al., 2007). In our sample (Karabekiroglu & Akbas, in press), 53.2% of the children with PDD-NOS had at least one co-morbid psychiatric disorder, including disruptive behavior disorders (40.4%), and anxiety disorders (18.0%). With respect to the PDD-NOS group, the ADHD group had significantly higher rates of co-morbid disruptive behavior disorders, learning disabilities, tic disorders, elimination disorders, and social anxiety disorder. On the other hand, the PDD-NOS group had significantly higher rates of co-morbid obsessive compulsive disorder with respect to the ADHD group. In a previous study, DeBurin et al. (2007) explored the comorbidity in ninety-four children with PDD-NOS, aged 6-12 years. At least one co-morbid psychiatric disorder was present in 80.9% of the children; 61.7% had a co-morbid disruptive behavior disorder, and 55.3% fulfilled criteria of an anxiety disorder. Compared to those without co-morbid psychiatric disorders, children with a co-morbid disorder had more deficits in social communication.

5. Conclusion

The overall results suggest that children with PDD-NOS have a high number of common features with patients having autism and ADHD. The symptoms of all three diagnostic groups appeared to form three clusters, "autistic spectrum," "ADHD spectrum," and "anxiety spectrum." Many features including language and motor development, "presenting" and/or "reported" symptom distribution, and gender distribution were found to be similar in the PDD-NOS and the autism groups. Mental retardation rate and symptom severity (e.g., "poor social interaction", "lack of eye contact", "stereotypies") were significantly higher in the autism group with respect to the PDD-NOS group. In addition, most of the previous studies supported quantitative discrimination rather than assuming that PDD-NOS and autism are qualitatively discrete groups. Therefore, PDD-NOS may be assumed as a partial subtype of autism and that it lies on a continuum of social-communication skill deficits. On the other hand, some of the studies suggest that these individuals appear to have a distinct variant of autism that does not merely sit at the less severe end of the same continuum of symptoms. They emphasize that compared to other disorders in PDD category, the children diagnosed with PDD-NOS had comparably severe but more circumscribed social communication difficulties, with fewer non-social features of autism. Therefore, we still need to investigate further the clinical features of children with PDD-NOS that distinguish them from children with autism and other non-PDD conditions.

6. References

Allen, D.A., Steinberg, M., Dunn, M., Fein, D., Feinstein, C., Waterhouse, L., et al. (2001). Autistic disorder versus other pervasive developmental disorders in young children: same or different? *European Child and Adolescent Psychiatry*, 10(1):67-78

American Psychiatric Assossiation (APA) (1994). Diagnostic and Statistical Manual of Mental Disorders, Fourth Edition. Washington DC: American Psychiatric Assossiation.

Barkley, R.A. (1990), A critique of current diagnostic criteria for attention deficit hyperactivity disorder: clinical and research implications. *Journal of Developmental and Behavioral Pediatrics* 11:343–352

Baron-Cohen, S., Wheelwright, S., Skinner, R., Martin, J., & Clubley, E. (2001). The Autism-Spectrum Quotient (AQ): Evidence from Asperger Syndrome/high-functioning autism, males and females, scientists and mathematicians. *Journal of Autism and Developmental Disorders*, 31: 5–17

Bishop, S.L., Richler, J., & Lord, C. (2006). Association between restricted and repetitive behaviors and NVIQ in children with autism spectrum disorders. *Child Neuropsychology*, 12 (4-5): 247-67.

Bryson, S.A., Corrigan, S.K., McDonald, T.P., & Holmes, C. (2008). Characteristics of children with autism spectrum disorders who received services through community mental health centers. *Autism*. 12(1):65-82.

Buitelaar, J.K., Van der Gaag, R., Klin, A., & Volkmar, F. (1999), Exploring the boundaries of pervasive developmental disorder not otherwise specified: analysis of data from the *DSM-IV* autistic disorder field trial. *Journal of Autism and Developmental Disorders*, 29:33–43

Chakrabarti, S., & Fombonne, E. (2001), Pervasive developmental disorders in preschool children. *JAMA* 285:3093–3099

Cohen, D.J., & Volkmar, F.R. (2005). Autismo e disturbi generalizzati dello sviluppo Trad. It. Brescia, Vannini Editrice.

deBruin, E.I., Ferdinand, R.F., Meester, S., de Nijs, P.F. & Verheij, F. (2007). High rates of psychiatric co-morbidity in PDD-NOS. *Journal of Autism and Developmental Disorders*. 37(5):877-86

deBruin, E.I., Verheij, F., & Ferdinand, R.F. (2006). WISC-R subtest but no overall VIQ-PIQ difference in Dutch children with PDD-NOS. *Journal of Abnormal Child Psychology*, 34(2):263-71

Eaves, L.C., Ho, H.H., & Eaves, D.M., (1994). Subtypes of autism by cluster analysis. *Journal of Autism and Developmental Disorders*, 24:3-22

Filipek, P.A., Accardo, P.J., Baranek, G.T., Cook E.H., Jr, Dawson, G., Gordon, B., et al. (1999), The screening and diagnosis of autistic spectrum disorders. *Journal of Autism and Developmental Disorders* 29:439–484

Gökler, B., Unal, F., Pehlivanturk, B., Cengel-Kultur, E., Akdemir, D., & Taner, Y. (2004). Okul çağı çocukları için duygulanım bozuklukları ve şizofreni görüşme çizelgesi - şimdi ve yaşam boyu şekli-Türkçe uyarlamasının geçerlik ve güvenirliği. *Turkish J Child Adolesc Mental Health*, 11 (3):109- 116

Hus, V., Pickles, A., Cook, E. H., Risi, S., & Lord, C. (2007). Using the Autism Diagnostic Interview-Revised to increase phenotypic homogeneity in genetic studies of autism. *Biological Psychiatry*, 61, 438–448.

Jensen,V.K., Larrieu, J.A., & Mack, K.K. (1997), Differential diagnosis between attention-deficit/hyperactivity disorder and pervasive developmental disorder-not otherwise specified. *Clinical Pediatrics (Phila)* 36:555–561

Karabekiroglu, K. & Akbas, S. (in press). Identifying and Differentiating PDD-NOS: A Comparison with Autism and ADHD. *New Symposium*.

Kamp-Becker, I., Ghahreman, M., Smidt, J., & Remschmidt, H. (2009). Dimensional Structure of the Autism Phenotype: Relations Between Early Development and Current Presentation. *Journal of Autism and Developmental Disorders*, 39:557–571

Kleinman, J.M., Ventola, P.E., Pandey, J., Verbalis, A.D., Barton, M., Hodgson, S., et al. (2008). Diagnostic Stability in Very Young Children with Autism Spectrum Disorders. *Journal of Autism and Developmental Disorders*, 38:606–615

Lord, C., Risi, S., DiLavore, P., Shulman, C., Thurm, A., & Pickles, A., (2006). Autism from two to nine, *Archives of General Psychiatry*, 63(6): 694–701

Luyster, R., Richler, J., Risi, S., Hsu, W.L., Dawson, G., Bernier, R., et al. (2005). Early regression in social communication in autistic spectrum disorders: a CPEA study. *Developmental Neuropsychology*, 27(3):311-336

Mandy, W., Charman, T., Gilmour, J., & Skuse, D., (2011). Toward Specifying Pervasive Developmental Disorder—Not Otherwise Specified, *Autism Research* 4: 1–11, 2011

Matson, J.L., Wilkins, J., Smith, K., & Ancona, M. (2007). PDD-NOS Symptoms in Adults with Intellectual Disability: Toward an Empirically Oriented Diagnostic Model. *Journal of Autism and Developmental Disorders*, 38(3):530-537.

Pearson, D.A., Loveland, K.A., Lachar, D., Lane, D.M., Reddoch, S.L., Mansour, R., et al. (2006). A comparison of behavioral and emotional functioning in children and adolescents with Autistic Disorder and PDD-NOS. *Child Neuropsychology*. 12(4-5):321-33

Prior, M., Eisenmajer, R., Leekam, S., Wing, L., Gould, J., Ong, B., et al.(1998). Are there subgroups within the autistic spectrum? A cluster analysis of a group of children with autistic spectrum disorders. *Journal of Child Psychology and Psychiatry*, 39:893-902

Rondeau, E., Klein, L.S., Masse, A., Bodeau, N., Cohen, D., Guile, J.M. (2010). Is Pervasive Developmental Disorder Not Otherwise Specified Less Stable Than Autistic Disorder? A Meta-Analysis. *Journal of Autism and Developmental Disorders*, Dec 14 (online)

Scheirs, J.G.M., Timmers, E.A., (2009). Differentiating Among Children with PDD-NOS, ADHD, and those with a Combined Diagnosis on the Basis of WISC-III Profiles. Journal of Autism and Developmental Disorders, 39:549–556

Sevin, J.A., Matson, J.L., Coe, D., Love, S.R., Mateses, M., & Benavidez, D.A. (1995). Empirically derived subtypes of pervasive developmental disorder. *Journal of Autism and Developmental Disorders*, 25:561–578

Simonoff, E., Pickles, A., Charman, T., Chandler, S., Loucas, T., & Baird, G. (2008). Psychiatric disorders in children with autism spectrum disorders: Prevalence, comorbidity, and associated factors in a population derived sample. *Journal of the American Academy of Child and Adolescent Psychiatry*, 47, 921–929.

Snow, A.V., & Lecavalier, L., (2011). Comparing Autism, PDD-NOS, and Other Developmental Disabilities on Parent-Reported Behavior Problems: Little Evidence for ASD Subtype Validity. *Journal of Autism and Develeopmental Disorders*, 41:302–310

Szatmari, P., Mérette, C., Bryson, S.E., Thivierge, J., Marc-Andre, R., Cayer, M., et al. (2002). Quantifying Dimensions in Autism: A Factor-Analytic Study. *Journal of American Academy of Child and Adolescent Psychiatry*, 41(4):467-474.

Tadevosyan-Layfer, O., Dowd, M., Mankoski, R., Winklosky, B., Putnam, S., McGrath, L., et al. (2003). A principal components analysis of the Autism Diagnostic Interview-Revised. *Journal of American Academy of Child and Adolescent Psychiatry*, 42:864–872.

Tanguay, P.E. (2004). Commentary: Categorical Versus Spectrum Approaches to Classification in Pervasive Developmental Disorders. *Journal of American Academy of Child and Adolescent Psychiatry*, 43(2):181-182.

Turner, L.M., Stone, W.L., Pozdol, S.L., & Coonrod, E.E. (2006). Follow-up of children with autism spectrum disorders from age 2 to age 9. *Autism*; 10(3): 243-65.

Van Lang, N. D. J., Boomsma, A., Sytema, S., de Bildt, A. A. Kraijer, D. W., Ketelaars, C., et al. (2006). Structural equation analysis of a hypothesized symptom in the autism spectrum. *Journal of Child Psychology and Psychiatry and Allied Disciplines*, 47(1), 37–44

Volkmar, F.R., Szatmari, P., & Sparrow, S.S.J. (1993). Sex differences in pervasive developmental disorders. *Journal of Autism and Developmental Disorders*, 23(4):579-91.

Volkmar, F. R., Lord, C., Bailey, A., Schultz, R. T., & Klin, A. (2004). Autism and pervasive developmental disorders. *Journal of Child Psychology and Psychiatry and Allied Disciplines*, 45(1), 135–170.

Wakabayashi, A., Baron-Cohen, S., Uchiyama, T., Yoshida, Y., Tojo, Y., Kuroda, M., et al. (2007). The autism-spectrum quotient (AQ) children's version in Japan: a cross-cultural comparison. *Journal of Autism and Developmental Disorders*, 37(3):491-500.

Walker, D.R., Thompson, A., Zwaigenbaum, L., Goldberg, J., Bryson, S.E., Mahoney, W.J., et al. (2004). Specifying PDD-NOS: A Comparison of PDD-NOS, Asperger Syndrome, and Autism. *Journal of American Academy of Child and Adolescent Psychiatry*, 43:172-180.

Waterhouse, L., Morris, R., Allen, D., Dunn, M., Fein, D., Feinstein, C., et al. (1996). Diagnosis and classification in autism. *Journal of Autism and Developmental Disorders*, 26:59-86.

Wing, L., & Gould, J. (1979). Severe impairments of social interaction and associated abnormalities, *Journal of Autism and Developmental Disorders*. 9:11–29

World Health Organization (WHO) (1993), *The International Classification of Diseases and Disorders-10 (ICD-10) Classification of Mental and Behavioural Disorders: Diagnostic Criteria for Research.* Geneva: World Health Organization

3

Autism and Genetic Syndromes

Willem Verhoeven, Jos Egger and Ilse Feenstra
Erasmus University Medical Centre, Rotterdam
Radboud University Nijmegen, Nijmegen
Vincent van Gogh Institute, Venray
The Netherlands

1. Introduction

Autism is a developmental disorder defined as a severe and persistent restriction in communicative skills, including lack of social and emotional reciprocity, as well as stereotyped and repetitive behaviours. Such an impairment of social interaction was already described in 1919 by the Swiss psychiatrist Eugen Bleuler within the framework of the negative symptom complex of schizophrenia. In the following decades in particularly the German language areas of Switzerland, autism was viewed by Kretschmer (1921) as a schizoid temperament whereas later, he viewed it as a special form of schizophrenia. In the late fifties, Leonhard (1957) assigned this specific disturbance in communication to the so called systematic schizophrenias. In 1943, the immigrated American child psychiatrist from Austria, Leo Kanner, originally described early infantile autism as Autistic Disturbance of Affective Contact. In a study of 11 children, four behavioural characteristics were distinguished: severe social withdrawal behavior, obsessive desire for repetitiveness, persistent fascination with specific objects or thoughts, and severe language impairments. One year later, the Austrian pediatrician Hans Asperger reported comparable findings under the title 'Die 'Autistischen Psychopathen' in Kindesalter'. Both Kanner (1943, 1971) and Asperger (1944) considered autism a communication disorder for children with severely impoverished relations with the environment (i.e., 'autistic aloneness').

Up until the beginning of the 1960s, under the influence of the then prevalent psychodynamic theories, autism was largely attributed to family and environmental factors. Rutter (1968) placed autism in a different perspective and demarcated the phenotypical presentation of both early infantile autism and schizophrenia from their biological underpinnings. Lorna Wing (1981) can be credited for bringing the descriptions of Asperger from 1944 back to our attention in the 1980s and, on the basis of extensive childhood epidemiological research, for placing autism in a broader diagnostic context and developing diagnostic criteria (Wing & Gould, 1979). Wing introduced the term 'autism spectrum disorder', which can be described on the basis of information from three domains: (a) social reciprocity, (b) verbal and non-verbal communication and imagination, and (c) a restricted, stereotyped pattern of interests and activities. These elements still constitute the diagnostic criteria from e.g. the DSM-IV category of Pervasive Developmental Disorders that include Autistic Disorder, Rett's Disorder, Childhood Disintegrative Disorder, Asperger's Disorder, and Pervasive Developmental Disorder Not Otherwise Specified (overview: Kumbier et al. 2010).

In her retrospective 'Reflections on opening Pandora's box' presented a few years back, Lorna Wing (2005) warns about stretching the boundaries of the autistic spectrum, which presently includes those who have normal to extremely high intelligence at the one end and those with a severe intellectual disability and limited social and communicative skills at the other end. In such a manner, the diagnostic label of 'autistic spectrum disorder' can possibly be misused to attain care (Volkmar et al., 2009). This is certainly not inconceivable in light of the quadrupled prevalence of pervasive developmental disorders across a period of 40 years (4.1 per 10,000 to 16.8 per 10,000), including somatic/neurological disorders that accompany autism (Fombonne, 2003; Rice, 2009). Moreover, autism and Asperger's disorder are regularly associated with other syndromes (Gillberg & Gillburg, 1989; Gillberg and Billstedt, 2000). As elegantly stated by Gillberg already in 1991 in his Emanuel Miller Memorial Lecture, researchers and clinicians, rather than allow themselves to be guided by stereotypic (sub)classifications, should be guided by a more balanced view of autism as belonging to a group of empathy disorders and thereby start the search for autism-relevant endophenotypes (Gillberg, 1992).

In the following sections, a brief overview of the current neuropsychological and genetic explanatory models will first be given. Thereafter, a selection of syndromes with a known genetic etiology that are seen to accompany autism will be discussed. To conclude, the manner in which modern genetics can be utilized in daily diagnostic practice will be elucidated.

2. Neuropsychological models

Roughly three explanatory models can be discerned for the pattern of disorders encountered in cases of autism. The *first* model is based upon the *weak central coherence* (WCC) hypothesis, which claims that patients with autism are inclined to attend to details as opposed to the whole during the processing of context-based information (i.e., strictly feature-based perception). As a consequence, the information is not understand as a meaningful whole and thus remains fragmented and confusing (Happé & Frith, 2006; Frith, 1989).

The *second* model draws upon the concept of *Theory of Mind*, which refers to the capacity to attribute thoughts, desires, and intentions to others. This capacity starts to develop around the age of three to four years and allows humans to take the perspectives of others into consideration in their own thinking. Deficiencies in this domain can easily lead to social misinterpretations and socially inadequate behavior, which form the basis of the severe communication problems that people with autism show (Baron-Cohen, 1989). In this connection, the more general notion of social cognition may be called upon and a link made to, for example, diminished activity of the amygdala and deviant perceptions of one's own emotions (Kethrapal, 2008).

In the *last* model, that of the *dysexecutive functioning* hypothesis, disturbed executive functions are assumed to play an important role. The executive functions (EF) are of major importance for the integration, steering, and control of processes required to execute purposeful behavior in new or complex situations. At a structural level, four frontal-subcortical circuits are involved in EF. The dorsolateral-prefrontal circuit has been related to executive cognitive dysfunctioning; the ventromedial circuit has been related to activation and motivation problems; and the medial- and lateral-orbitofrontal circuits have been related to disturbed affect regulation and disturbed social behavior, respectively (Chow &

Cummings, 2007; Alvarez & Emory, 2006). EF often involves the overruling of automatic responses in favour of more intentional behaviour. A capacity to flexibly switch between different behavioral repertoires (i.e., *monitoring* and *shifting*) is required for such overruling and typically found to be a problem in cases of autism. Barnes-Holmes and colleagues view EF as rule-governed behaviour and thus behaviour that stands in contrast to contingency-shaped behaviour, which has been automatized. EF thus defined, is verbal behaviour that precedes other behaviour (i.e., *verbal antecedent behavior*) and therefore distracts the individual from his usual, automatic reaction pattern. Stated differently, the probability of an alternative behavior is changed in the direction of a particular objective. In such a manner, thus, recent research on rule-governed behaviour connects EF with autism and an underlying Theory of Mind (Barnes-Holmes et al., 2004; McHugh et al, 2004).

3. Genetic models

Autism can be viewed as a classic example of a disorder with a strong genetic basis. A distinction must nevertheless be made between the Autistic disorder as originally described by Kanner and the autism spectrum disorder, which can be viewed as a component of an array of clinical pictures and syndromes that are sometimes referred to as secondary or syndromic autism (Benvenuto et al., 2009). Given the complex interplay between genes and autism, a search for at least two types of genetic factors is of importance, namely: (rare) chromosomal abnormalities or gene alterations that can be directly related to core (i.e., classic) autism and genetic copy number variants that correlate with a vulnerability to develop an autistic disorder.

In several studies from the 1980s and 1990s that use a strict definition of autism, a 69% to 95% concordance has been demonstrated in monozygotic twins, while the chance in dizygotic twins is only 0% to 24%. The contribution of the hereditary components is estimated to be 90%. The male-female ratio is between 3-4 to 1 (Brkanac et al., 2008). In order to advance the understanding of the genetic heterogeneity of autistic disorders, various techniques can be used such as (molecular) cytogenetic research, linkage studies, and association studies.

Linkage studies search for those parts of a chromosome that are found to be the same for all affected individuals in a family but different for the non-affected family members. A gene that contributes to the occurrence of a vulnerability for autism may lie in such a shared region. These studies have revealed a wide variety of loci, from which a considerable genetic heterogeneity can be concluded as well as the absence of single, specific locations for autism (Szatmari et al., 2007). Recently, in a very large-scale linkage study, two new locations have been found on chromosomes 6 and 20 (6q27 and 20p13, respectively) for which the functional significance has yet to be clarified (Weiss et al., 2009).

The same holds for candidate genes that have been implicated in a large series of association studies. These studies investigate significant genetic differences between large groups of patients, on the one hand, and groups of healthy individuals, on the other hand (Vorstman et al., 2006a). The research findings make it clear, however, that the pathophysiology of autism involves genes that code for proteins from the family of neurexins and neuroligins that play, in turn, a role in the development and functioning of synaptic and in particular glutamatergic and GABA-ergic networks (Lisé & El-Husseini, 2006; Buchsbaum, 2009). The first X-linked mutations in genes involved in the coding of neuroligin were revealed in patients with autism in two Swedish families in 2003 (Jamin et al., 2003).

When the classic microscopic cytogenetic route is followed, structural chromosomal aberrations are found in 3% to 7% of patients with autism and developmental delay. This finding concerns mainly maternal duplications on the long arm of chromosome 15 (q11-13) (Hogart et al., 2010) and deletions on the long arm of chromosome 2 (q37) (Falk & Casas, 2007), chromosome 7 (q22 and q31) (Alarcón et al., 2002) and chromosome 22 (q11) (Niklasson at al., 2009) and (q13) (overview: Kumar & Christian, 2009). The fluorescence-in-situ-hybridization (FISH) technique is used to search for specific submicroscopic deletions and is used primarily to confirm a clinical diagnosis such as the 22q11 deletion syndrome. Disadvantages of this technique are its labor intensiveness and that only one or a few chromosome regions can be examined per experiment.

More commonly used these days is the whole-genome microarray technique. Here, details more than a hundred times smaller can be perceived when compared to microscopic cytogenetic examination (de Vries et al., 2005; Veltman & de Vries, 2006). With the aid of such a 'DNA chip', the complete genome with a high resolution can be examined for the presence of microdeletions and duplications or so-called copy-number variations (CNVs). Of principal concern here are small quantitative, structural variations that are paired with a loss or gain of chromosome material. Furthermore, the 'targeted genomic array' technique can be applied to study specific regions such as the subtelomeric chromosome regions or well-known microdeletion syndrome regions (Lintas & Persico, 2009).

For various neuropsychiatric disorders including autism, CNVs possibly involved in the vulnerability for the development of a disorder within the autism spectrum have been demonstrated using the array technique (Jacquemont et al., 2006; Cook & Scherer, 2008). For some of these CNVs, a clear correlation has been demonstrated, e.g., a maternal 15q11-q13 duplication was shown for 1-3% of patients with an autistic spectrum disorder. Another frequently occurring CNV is found on the chromosome 16p11.2 which present with a deletion or duplication in approximately 1% of the patients (Weiss et al., 2008; Fernandez et al., 2010).

There are, however, also CNVs found with an, as yet, unknown significance; because, for example, the same change can be traced back to a healthy parent. A precise interpretation of the array results with the aid of bio-informatics, literature databases, data from the pedigree and clinical investigation of affected individuals, is therefore essential.

Another interesting perspective is the neuropeptide concept. It has been known for quite some time that the nonapeptide oxytocine (OXT) is involved in affiliation behaviour and social cognition via an improvement of social memory, including the recall and understanding of affective events (Hollander et al., 2007; Insel, 2010; Green & Hollander, 2010). For these reasons, research has been performed on the association between single nucleotide polymorphisms (SNPs) in the OXT gene and the OXT receptor gene (OXTr). Relative to the normal population, more SNPs were present in the OXTr for a subgroup of patients with autism, which could indicate a genetically determined vulnerability for the development of autism (Lerer et al., 2008; Lee et al., 2009). In line with these observations, Gregory et al. (2009) found that epigenetic regulation of OXTr is implicated in the development of autism.

To summarize, in linkage and association studies among patients with autistic disorders up until now, a large number of candidate genes and gene locations have been found for which it can be assumed that they may be involved in the development of functional processes of the central nervous system. In Table 1, a selective overview is presented. In the following section, the most well-known genetic disorders associated with autism will be discussed.

Chromosome	*Candidate Gene*	*Network/Function*
2p16.3	NRXN1	Synapse formation
2q12.3-q14.2	DPP10	Neurotransmission
3p24-26	OXTr	Neurotransmission
3p26-p25	CNTN4	Synapse formation
4p14-q21.1	GABRG/GABRA	GABA neurotransmission
7q31.1	ST7	Tumor suppression
7q35-q36	CNTNAP2	Synapse formation
8p23	DLGAP2	NMDA neurotransmission
15q11-q14	GABRA/GABRB/GABRG	GABA neurotransmission
15q13	APBA2	Neurotransmission
16p11.2	DOC2A	Neurotransmission
22q11	PRODH	Neuromodulation
22q13	SHANK3	Synapse formation
Xp22.3	NLGN4	Synapse formation
Xp11.4	TSPAN7	Neuronal growth and development
Xp22.1-p21.3	IL1RAPL1	Interleukin receptor

Table 1. Selection of genes and functions possibly involved in autism (adapted from Guilmatre et al., 2009 and El-Fishawy & State, 2010)

4. Genetic syndromes and autism

4.1 Fragile X syndrome

The fragile X syndrome (FXS; Figure 1) is the most well-known genetic disorder related to autism. Brown and colleagues (1982) were the first to report on this. Initially, FXS was described by Lubs (1969), who detected a fragile site at the end of the long arm of the X chromosome by using classical microscopic cytogenetic techniques. FXS is caused by hypermethylation of an expanded trinucleotide repeat (CGG) in the 'fragile X mental retardation 1 (FMR1) gene' (Xq27.3). In normal individuals, the number of CGG repeats is 5 to 45 which is stably transmitted to the next generation. In case of a FMR1 premutation, there is a small expansion of 55 to 200 repeats. In FXS, the number of repeats exceeds 200. As a result of this enlarged number of repeats, hypermethylation of the FMR1 gene occurs which leads, in turn, to a shortage or complete loss of the FMR1 protein that is essential for dendrite formation, synapse formation, and experiential learning (Marco & Skuse, 2006; Hernandez et al., 2009).

FXS is an X-linked disorder with an incidence of about 1 in 4000 newborn males. Affected males show a variable degree of developmental delay, behaviour problems, and distinctive dysmorphic features such as a long face and large, prominent ears. Female carriers with a full mutation (>200 repeats) may present with or without impaired level of intelligence. Females with FXS and normal intelligence, however, have an increased risk of mood and anxiety disorders and a schizotypical personality disorder (Franke et al., 1998).

In males with a premutation (50-200 repeats), there is an increased probability of the development of the so-called fragile-X-associated tremor/ataxia syndrome (FXTAS). Its symptoms emerge at a later age, and the syndrome has a progressive course. The clinical picture comprises intention tremor, frequent falling, Parkinsonian symptoms, and disturbed cognitive and executive functioning (Verhoeven et al., 2008; Bourgeois et al., 2009).

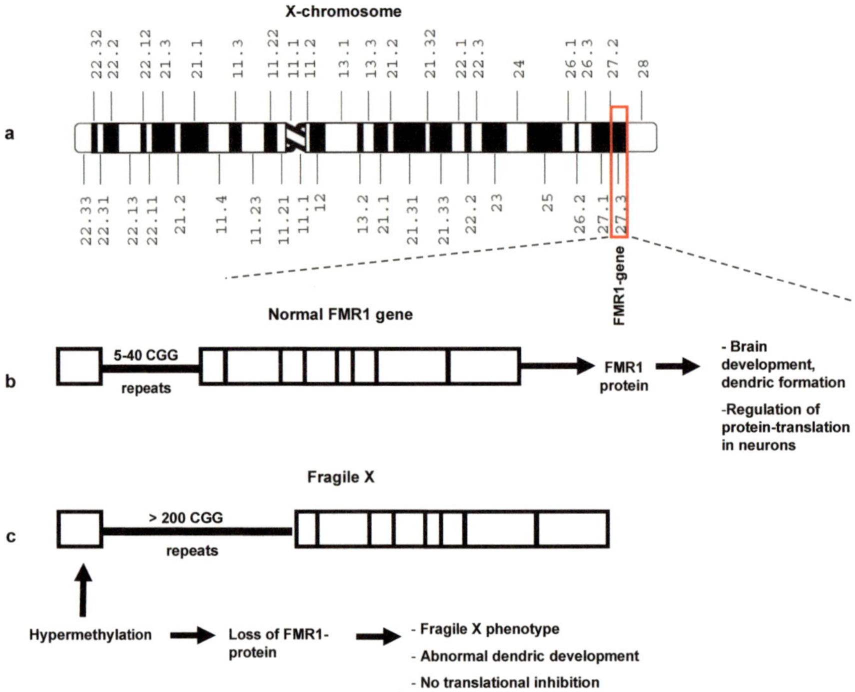

Fig. 1. Fragile X syndrome and the FMR1 gene

Schematic representation of the location of the FMR1 gene on the long arm of the X chromosome. (a) The FMR1 gene comprises a polymorph repetition of the base pairs cytosine-guanine-guanine (CGG) at the start of the gene. In healthy individuals, this number of CGG repeats varies from 5 to about 45 units. (b) Affected individuals have more than 200 CGG repeats: the full mutation. (c) The expansion to a full mutation (>200 repeats is usually associated with hypermethylation of the CGG repetition and the adjacent area, which leads to a transcription stop resulting in the absence of the FMR1 protein. This results in intellectual disability and other symptoms of the Fragile X syndrome among men, and in >50% of the women who are carriers of a full mutation. Carriers with 55 and 200 repeats are asymptomatic and are thus called premutation carriers

For decades, it has been known that the severity of intellectual disability and the intensity of related behaviour problems is proportional to the number of repeats. The psychopathological phenotype of FXS includes, in addition to the developmental delay, multiform anxiety symptoms, obsessive-compulsive characteristics, hyperactivity / impulsivity, and aggression. Epileptic phenomena are frequent. Predominant, however, are autism-related symptoms such as social anxiety and withdrawal behaviour, stereotypies like flapping or biting of the hands, perseverations, extreme sensitivity to environmental stimuli, and, in general, decreased social reciprocity with an avoidance of eye contact (Hagerman, 2005).

Using neuroimaging techniques, various structural abnormalities of the central nervous system can be demonstrated, in particular enlargement of hippocampus, amygdala, caudatus, and thalamus with a reduction of the cerebellar vermis (Hessl et al., 2004). These neuronal changes are caused by an overabundance of immature dendritic spines. In normal conditions, dendritic spines are essential for the formation of new neuronal connections that, in turn, form the basis for learning and memory. In FXS, the cognitive dysfunctions in the domains of attention, (working) memory, mathematical skills, executive functioning and social cognition largely correspond to the observed abnormalities of the central nervous system.

The treatment of patients with FXS is primarily symptomatic and aimed at the reduction of the most prominent behavioural problems or psychiatric symptoms, such as anxiety, hyperactivity, impulsivity and distractibility (Garber et al., 2008). Since the extremely heightened sensitivity to environmental stimuli is assumed to underlie the above mentioned symptoms, it is essential to reduce excessive environmental sensory activation. This can be achieved with a more structured daily program of activities, the promotion of a realistic pattern of expectations among parents/caregivers, individualized instruction and, most importantly, the dissemination of knowledge about the persistence of the FXS behavioral phenotype.

4.2 Rett syndrome

Rett syndrome (RS) was first described in 1966 by Andreas Rett. This syndrome is inherited in an X-linked manner, caused by a mutation in the Methyl-CpG-Binding Protein 2 (MECP2) gene (Xq28). In Figure 2, the location of the MECP2 gene is depicted. RS occurs almost exclusively in girls and its prevalence is estimated to be between 1/10,000 - 1/20,000. In boys, the disorder is nearly always lethal. In rare male cases, an extra X chromosome or mosaicism of the MECP2 mutation has been found.

RS is characterized by an apparently normal development in the first 6 to 18 months of life after which development stagnates, acquired skills get lost and development finally stops. This stagnation of development also becomes manifest in a retarded and disproportionate growth in head circumference, decrease of eye contact, and both cognitive and motor deterioration. Already in the first year of life, autistic behavioural elements are present such as social withdrawal, declining speech and communication, limited eye contact, grinding of the teeth, and characteristic hand stereotypies (Ben Zeev, 2007; Gonzales & LaSalle, 2010). In the majority of the patients, the syndrome is associated with severe epilepsy. The first decade is dominated by severe neurological dysfunctions and an irregular respiration pattern as a result of an immaturely developed brainstem. In addition, a prolonged QT interval is often present with, as a consequence, risk for sudden cardiac arrhythmia. From the age of 10, a developmental plateau occurs and the patient becomes severely neurologically handicapped with profound intellectual disability.

In 80% to 90% of the patients with RS, a mutation that almost always occurs *de novo*, can be demonstrated in the MECP2 gene. This gene is expressed particularly in neurons and to a lesser extent in glial cells, and involved in neuronal maturation in the postnatal period. MECP2 is involved in the expression of the gene that codes for brain derived neurotrophic factor (BDNF), which is essential for neuronal maturation and plasticity. The pathophysiology of RS thus lies conceivably in a MECP2-mediated disturbance in the regulation of BDNF. It is assumed that the severity of the disorder and the progression over time of the successive stages corresponds with a polymorphism in BDNF (Matijevic et al., 2009; Ben Zeev et al., 2009).

RS is one of the better examples of an autism-related disorder with a proven genetic pathophysiology. While there are clear differences between classical autism and the phenotypic presentation of autism in RS (i.e., characteristic stereotypy such as hand-wringing at chest level and a relative maintenance of eye contact), research in RS could nevertheless contribute to a better understanding of involvement of central nervous system dysfunctions in autism.

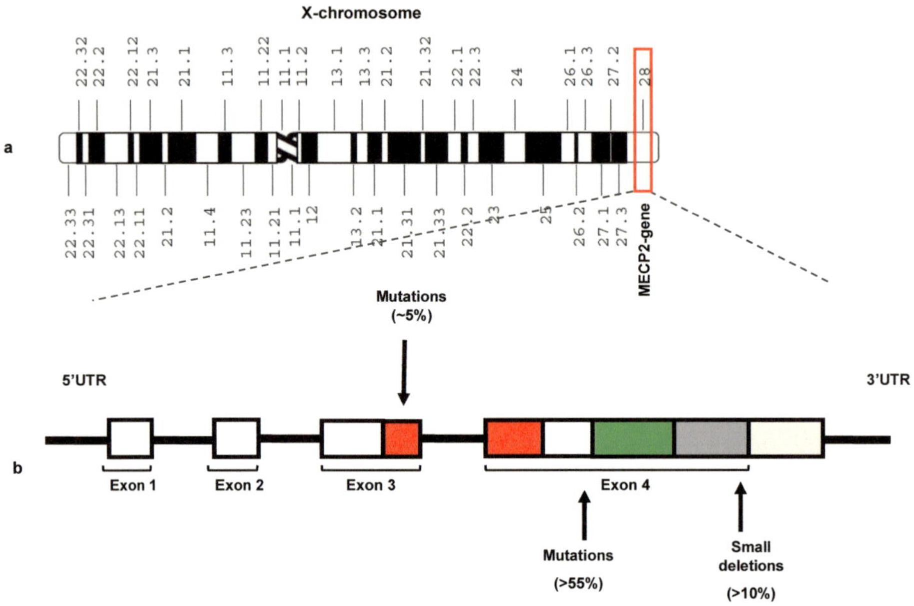

Fig. 2. Rett syndrome and the MECP2 Gene
Schematic representation of the location of the MECP2 gene on the long arm of the X chromosome. (a) The MECP2 gene is constituted by 4 coding exons.The majority of mutations among patients with Rett syndrome are found in exon 4. In addition, in more than 10% of the patients, a deletion of the last part of exon 4 is present. The terms 5' UTR (UnTranslated Region) and 3' UTR indicate the direction in which the gene is read; from 5' UTR to 3' UTR

4.3 Tuberous sclerosis

Tuberous sclerosis complex (TSC) of Bourneville-Pringle was first described in 1880 by Bourneville and is a multi-organ disorder with a autosomal dominant inheritance. The prevalence is 1 in the 6,000 - 10,000 births. TSC is caused by mutations in two genes, the TSC 1 and 2 gene. The TSC1 gene is located on chromosome 9 (9q34.3) and codes for hamartin while the TSC2 gene is located on chromosome 16 (16p13.3) and codes for tuberin. In approximately 85% of patients with a clinically confirmed diagnosis of TSC, a change in one of these two genes can be demonstrated. Usually, a *de novo* mutation is present, although 30% of the index patients has one or more affected family members. Mutations in both genes can lead to abnormal cell growth and differentiation in multiple organ systems. In the brain, this is expressed by the formation of cortical and subcortical hamartomas including tubers. In addition, various organ systems can be affected leading to the development of cystic kidneys, angiofibromas of the face, and rhabdomyomas. The structural abnormalities of the central nervous system are associated with various forms of epilepsy, cognitive dysfunctions and symptoms of autism (Datta et al., 2008). There is, however, a great variability in the severity of the clinical characteristics across TSC patients, also within one single family.

Already in 1932 and thus *before* Kanner's publication, Critshley and Earl described the autistic characteristics associated with TSC, being decreased social contact, stereotypies, disturbed speech, and withdrawal behaviour. Research during the last few decades has shown autism to occur in about 25 to 60 percent of TSC patients, although its symptom profile differs qualitatively from that seen in classical autism. A higher level of social-cognitive functioning as well as less pronounced stereotypies are characteristic of patients with TSC. Moreover, in contrast to autism, the male-female ratio in TCS is about equal (Wiznitzer, 2004).

The neurobiological substrate for autism in TSC is still unclear. For both hamartin and tuberin, it is assumed that both proteins modulate cell function and play a role in neuronal migration, differentiation, and development and that they together form a functional complex (Asato et al., 2004). The latter can be considered as a type of 'neuronal polarity' in which over expression of the TSC1/TSC2 complex suppresses the formation of axons while an under expression is associated with the formation of tubers (Choi et al., 2008). This TSC1/TSC2 functional integration may explain that a mutation in one of the two genes can result in the same phenotype (Orlova & Crino, 2010). Finally, it has been demonstrated that the number of tubers in the brain correlates with the incidence of autism and that their localization corresponds with the type of epilepsy (Marcotte & Crino, 2006).

In sum, also for TSC, it is clear that the presence of autistic behavioural characteristics relates to a well-defined gene defect. This knowledge from TSC research may further elucidate the pathophysiology of autism.

4.4 22q11 deletion syndrome

The 22q11 deletion syndrome (22q11DS) was first described in 1978 by Shprintzen as velo-cardio-facial syndrome and is caused by an interstitial deletion of chromosome 22 (22q11.2). In Figure 3, a micro-array profile of chromosome 22 from a patient with 22q11DS is depicted.

This syndrome is associated with, among others, congenital heart and conotruncal defects, cleft palate, hypoparathyroidism, and facial dysmorphisms. The prevalence of 22q11DS is 1:4,000 with an equal male-female distribution. The deletion involved in this syndrome can encompass multiple genes, with the T-box 1 (TBX1) gene as the most important. Its encoded protein is crucial for the development of specific brain areas, heart, face, and limbs (Paylor et al., 2006). It is, however, doubtful whether this gene also plays a role in the etiology of psychiatric disorders that often accompany 22q11DS (Funke et al., 2007).

During the past decades, it has become clear that psychiatric disorders often occur in 22q11DS patients. These include psychoses in particular (Vogels et al., 2002; van Amelsvoort et al., 2004; Verhoeven et al., 2007), but also anxiety, mood, and obsessive-compulsive disorders (Shprintzen, 2000). In addition, in 15% to 30% of the patients with 22q11DS, autistic features are present such as withdrawal behaviour, impaired social interaction, reduced facial expression, and cognitive deficits, e.g., perseveration, reduced mental flexibility, and restricted problem-solving capacities (Fine et al., 2005; Vorstman et al., 2006b; Anshel et al., 2007; Niclasson et al., 2009). Closer inspection of the psychopathological profile has demonstrated that both the psychotic and the autistic symptoms evolve from a diminished comprehension of abstract and symbolic language, in addition to a limited capacity to correctly estimate the intentions, emotions, and behaviour of others (Sphrintzen, 2000; Verhoeven et al., 2007).

In sum, for 22q11DS, it is obvious that detailed analysis of the cognitive, emotional, and psychiatric profile is of critical importance for the choice of an individual treatment strategy.

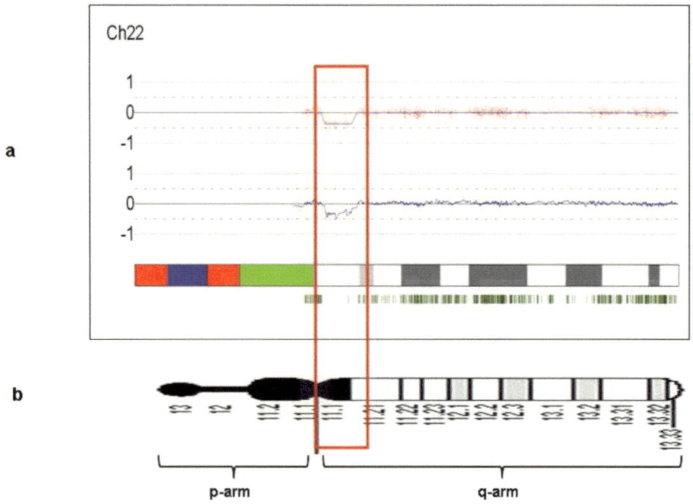

Fig. 3. Microarray profile of chromosome 22 from a patient with 22q11DS
(a) Representation of a microarray profile of chromosome 22 from a patient with a 22q11 deletion. (b) The ideogram of chromosome 22 with the indication of short p arm and the long q arm is depicted underneath.. In the upper portion, every individual clone is represented separately by a red dot on the X axis running between the end of the p arm (left) and the end of the q arm (right). On the Y axis, the amount of DNA from the patient as compared to that from control samples (CK) can be read. In case of an equal amount, the log2 ratio approximates zero. In cases of deletion, this will be -1 or lower. In cases of duplication, this will be +1 or higher. The p arm of chromosome 22 is not represented in the microarray profile because it only consists of satellites and non-coding material

4.5 Metabolic disorders

While genetically determined metabolic disorders are relatively rare, nevertheless, they often manifest with disorders along the autistic spectrum. The establishment of such a diagnosis is of importance for not only treatment and prognosis but also for gaining more insight into the pathophysiology of autism. Primarily involved are disturbances in amino acid metabolism such as phenylketonuria, disorders in purine metabolism, creatine deficiency syndromes, Smith-Lemli-Opitz syndrome (i.e., an inherited defect in the synthesis of cholesterol), urea cycle disorders, and mitochondrial disorders (Manzi et al., 2008; Zecavati & Spence, 2009; see Table 2). The latter may even have its debut with symptoms from an autism spectrum disorder (Weissman et al., 2008).
From the metabolic disorders, the creatine deficiency syndromes represent a recently recognized group of diseases that are caused by inherited defects in the biosynthesis and transport of creatine. Two defects in the biosynthesis have been reported that include deficiencies of the enzymes L-arginine-glycine amindinotransferase (AGAT) and guanidinoacetate methyltransferase (GAMT). The third is a functional defect involving the creatine transporter mechanism. The latter is an X-linked syndrome caused by a defective creatine transporter and was first described by Salomons et al. (2001). It appeared to be the

result of a mutation in the creatine transporter gene called SLC6A8 that was mapped to Xq28. Its prevalence is estimated to be at least 2% of X-linked mental retardation syndromes (Rosenberg et al., 2004). Since the SLC6A8 gene is expressed in most tissues (e.g. skeletal muscle, kidney, colon, brain and heart), several organ systems can be affected.

Disorder	*First appearance*	*Characteristics*	*Treatment option*
Phenylketonuria	Neonatal	Intellectual disability, autism, epilepsy	Special diet
Purine metabolism disorders	From childhood	Development delay, autistic characteristics, impulsivity, epilepsy	None
Creatine deficiency syndromes	3 months - 2 years	Autistic characteristics, epilepsy/myoclonic twitches, language and developmental delays, extrapyramidal symptoms	Creatine substitution
Cholesterol synthesis defects	From childhood	Autism, psychomotor retardation	Cholesterol suppletion
Urea cycle disorders	Neonatal and postnatal	Hyperactivity, epilepsy, intellectual disability, autism	Special diet measures
Sanfilippo syndrome	Variable, depending on subtype	Loss of acquired skills, autism	None
Mitochondrial disorders	Variable, depending on subtype	Depending on organ system	None

Table 2. Metabolic disorders and autism (adapted from Zecavati & Spence, 2009 and Manzi et al., 2008)

The main organ involved in creatine deficiency syndromes is the central nervous system. Patients show severe neurodevelopmental delay and, from early infancy on, mental retardation, epilepsy, disturbances in active and comprehensible speech, autism and self-injurious behaviour become prominent (Salomons et al., 2003; Béard and Braissant, 2010). In patients with GAMT or AGAT deficiency, early oral creatine substitution treatment might effectively prevent neurological sequelae. In those with a defect in the creatine transporter gene SLC6A8, however, suppletion with L-arginine is not effective at all (Nasrallah et al., 2009).

Although metabolic syndromes should always be involved in the differential diagnosis of autism spectrum disorder, systematic screening of such patients is only mandatory in case of a suggestive actual symptomatology and/or developmental history. An example is the Sanfilippo B syndrome, a mucopolysaccharidosis caused by a mutation in the NAGLU gene, which leads to an accumulation of heparan sulfate with, as a consequence, damage to the central nervous system and various organ systems. This diagnosis was recently determined for an older, mild intellectually disabled patient who was referred for behavior problems with a history of limited verbal and emotional communication, stereotypies, impulsivity, and anxieties (Verhoeven et al., 2010).

5. Closing remarks

The majority of patients with autism present with a mild to severe intellectual disability. In a substantial number of cases, moreover, the autistic disorder appears to be part of a genetic disorder. It is quite remarkable therefore, that *only one* genetic disorder from the array of genetic disorders associated with autism, is included in the DSM-category of autistic disorders, namely the neurodegenerative Rett syndrome. It should, however, be emphasized that the identification of autistic behaviours and the diagnosis of an autism spectrum disorder is extremely difficult in patients with severe intellectual disability in the context of a genetic syndrome (Moss & Howlin, 2009). It is also evident that in case of exceptionally high intelligence, Asperger's disorder or atypical autism are usually the autistic disorders involved. For this specific group of patients, however, no information on genetics is available as yet. These patients are often subsumed under the general DSM category 'Pervasive Developmental Disorder,Not Otherwise Specified'.

Apart from the changes of diagnostic concepts over the past decades, research on the genetic underpinnings of autism and related disorders confronts three major complexities. First, there is the large degree of genetic heterogeneity, which means that different genes can contribute in a varying way to the emergence of a disorder. A second difficulty is the polygenetic inheritance; that is, the simultaneous presence of multiple genetically-determined vulnerabilities that may be responsible for the development of a particular syndrome. A third problem lies in the well-known interaction between environmental and genetic factors during development from early conception on (Volkmar et al., 2009).

All mutations that are causative for the aforementioned disorders concern genes involved in the early development of the central nervous system. The search for susceptibility genes has made it clear that disturbed synaptic transmission in, for example, the neuroligin network is involved in the pathophysiology of a certain, albeit small, percentage of cases with autism. This kind of knowledge might be relevant for the development of putative future pharmacological treatment strategies for a subgroup of patients with autism. In this context, the earlier mentioned significance of the neuropeptide OXT could also be noteworthy.

The results from a large number of studies during the past decades lead to several conclusions. It is clear that autism, both phenotypically and genotypically, is a very heterogeneous disorder and that the quest for the grail of a single-high-impact gene will never succeed. In general, mutations or common variants in genes are thought to be involved in the neuronal domains, synaptic interaction, neurotransmission, and cell migration and growth (Freitag et al., 2010). Attention should therefore be shifted to large-scale screening for *de novo* mutations and CNVs that can influence the functioning of a gene (Sebat et al., 2007; Vissers et al., 2010).

Recently, all information available on the vulnerability genes and CNVs associated with autism has become available via the Autism Genetic Database (AGD), that can be freely accessed at http://wren.bcf.ku.edu (Matuszek & Talebizadeh, 2009). In addition, modern fMRI techniques may be of use to map neuronal endophenotypes that are critical for further genetic studies of autism (Losh et al., 2008; Piggot et al., 2009).

For daily clinical practice, facial dysmorphisms in patients with autism in addition to intellectual disabilities, constitute the initial indication for modern genetic investigation.

Epilepsy at young age and gradual deterioration of previously acquired skills warrant further search for a metabolic disorder. Future scientific studies may reveal to which extent sets of genes are involved in the pathophysiology of autism and autism related disorders per se, but also of neuropsychiatric disorders in general (Lichtenstein et a., 2010). In all cases it is clear that detailed information about developmental history, neuropsychiatric/ neuropsychological profile as well as an elaborative inventory of family characteristics is mandatory for appropriate genetic search. This holds true for both the individual patient and for a group of well defined patients.

6. References

Alarcón, M., Cantor, R.M., Liu, J., Giliam, T.C., The Autism Genetic Resource Exchange Consortium & Geschwind, D.H. (2002). Evidence for a language quantitative trait locus on chromosome 7q in multiplex autism families. *American Journal of Human Genetics*, 70, 60-71.

Alvarez, J. A., & Emory, E. (2006). Executive function and the Frontal Lobes: a meta-analytic review *Neuropsychology Review, 16*, 17-42.

Antshel, M., Aneja, A., Strunge, L., Peebles, J., Fremont, W.P., Stallone, K., et al. (2007). Autistic spectrum disorders in velo-cardio facial syndrome (22q11.2 deletion). *Journal of Autism and Developmental Disorders, 37*, 1776-1786.

Asato, M.R. & Hardan, A.Y. (2004). Neuropsychiatric problems in tuberous sclerosis complex. *Journal of Child Neurology, 19*, 241-249.

Asperger, H. (1944). Die "Autistischen Psychopathen" im Kindesalter. *Archiv für Psychiatrie und Nervenkrankheiten, 1*, 76-136.

Barnes-Holmes, Y., Barnes, D., McHugh, L., & Hayes, S. C. (2004). Relational Frame Theory: Some Implications for Understanding and Treating Human Psychopathology. *International Journal of Psychology and Psychological Therapy, 4*, 355-375.

Baron-Cohen, S. (1989). The autistic child's theory of mind : A case of specific developmental delay. *Journal of Child Psychology and Psychiatry, 30*, 285-297.

Béard, E. & Braissant, O. (2010) Synthesis and transport of creatine in the CNS: improtance for cerebral functions. *Journal of Neurochemistry*, 115, 297-313.

Benvenuto, A., Moavero, R., Alessandrelli, R., Manzi, B. & Curatolo, P. (2009). Syndromic autism: causes and pathogenetic pathways. *World Journal of Pediatrics, 5*, 169-176.

Ben Zeev Ghidoni, B. (2007). Rett syndrome. *Child and adolescent psychiatric clinics of North America, 16*, 723-743.

Ben Zeev, B., Bebbington, A., Ho, G., Leonard, H., de Klerk, N., Gak, E. et al. (2009). The common BDNF polymorphism may be a modifier of disease severity in Rett syndrome. *American Academy of Neurology, 72*, 1242-1247.

Bourgeois, J.A., Coffey, S.M., Rivera S.M., Hessl, D., Gane, L.W., Tassone F., et al. (2009). A review of fragile X premutation disorders: expanding the psychiatric perspective. *Journal of Clinical Psychiatry, 70*, 852-862.

Bourneville, D.M. (1880) Sclérose tubéreuze des circonvolutions cérébrales: idiote et épilipsie hémiplégique. *Archives de Neurologie, 1*, 81-91.

Brkanac, Z., Raskind, W.H. & Kin, B.H. (2008). Pharmacology and genetics of autism: implications for diagnosis and treatment. *NIH Public Access, 5*, 599-607.

Brown, W.T., Jenkins, E.C., Friedman, E., Brooks, J. Wisniewski, K., Raguthu, S., et al. (1982). Autism is associated with the Fragile-X syndrome. *Journal of Autism and Developmental Disorders, 12,* 303-308.

Buxbaum, J.D. (2009). Multiple rare variants in the etiology of autism spectrum disorders. *Dialogues in Clinical Neuroscience, 11,* 35-43.

Choi, Y.J., Di Nardo, A., Kramvis, I., Meikle, L., Kwiatkowski, D.J., Sahin, M., e.a. (2008). Tuberous sclerosis complex proteins control axon formation. *Genes & Development 22,* 2485-2495.

Chow, T.W. & Cummings, J.L. (2007). Fronto-subcortical circuits. In: B.L. Miller & J.L. Cummings (Eds.), *The human frontal lobes: Functions and disorders* (pp. 25–43). New York: Guilford Press.

Cook, E.H. & Scherer, W. (2008). Copy-number variations associated with neuropsychiatric conditions. *Nature, 455,* 919-923.

Critchley, M., Earl, C.J.C. (1932) Tuberous sclerosis and allied conditions. *Brain, 55,* 311-346.

Datta, A.N., Hahn, C.D. & Sahin, M. (2008).Clinical presentation and diagnosis of tuberous sclerosis complex in infancy. *Journal of Child Neurology, 23,* 268-273.

De Vries, B.B.A., Pfundt, R., Leisink, M., Koolen, D.A., Vissers, L.E.L.M., Janssen, I.M., et al. (2005) Diagnostic genome profiling in mental retardation. *American Journal of Human Genetics, 77,* 606-616.

El-Fishawy, P & State, M.W. (2010). The genetics of autism: key issues, recent findings and clinical implications. *The Psychiatric Clinics of North America, 33,* 83-105.

Falk, R.E. & Casas, K.A. (2007). Chromosome 2q37 deletion: clinical and molecular aspects. *American Journal of Medical Genetics, 145C,* 357-371.

Fernandez, B.A., Roberts, W., Chung, B., Weksberg, R., Meyn, S., Sztmari, P., e.a. (2010). Phenotypic spectrum associated with de novo and inherited deletions and duplications at 16p11.2 in individuals ascertained for diagnosis of autism spectrum disorder. *Journal of Medical Genetics, 47,* 195-203.

Fine, S.E., Weissman, A., Gerdes, M., Pinto-Martin, J., Zackai, E.M., McDonald-McGinn, D.M., Emauel, B.S. (2005). Autism spectrum disorders and symptoms in children with molecularly confirmed 22q11 deletion syndrome. *Journal of Autism and Developmental Disorders 35,* 461-470.

Fombonne, E. (2003). Epidemiological surveys of autism and other pervasive developmental disorders: an update. *Journal of Autism and Developmental Disorders, 33,* 365-382.

Franke, P., Leboyer, M., Gänsicke, M., Weiffenbach, O., Biancalana, V., Cornillet-Lefebre, P., et al. (1998). Genotype-phenotype relationship in female carriers of the premutation and full mutation of FMR-1. *Psychiatry Research, 80,* 113-127.

Freitag, C.M., Staal, W., Klauck, S.M., Duketes, E. & Waltes, R. (2010) Genetics of autistic disorders: review and clinical implications. *European Child & Adolescent Psychiatry, 19,* 169-178.

Frith, U. (1989). *Explaining the enigma.* Oxford: Blackwell

Funke, B.H., Lencz,T., Finn, C.T., DeRosse, P., Poznik, G.D., Plocik, A.M. et al. (2007). Analysis of TBX1 variation in patients with psychotic and affective disorders. *Molecular Medicine, 13,* 407-414.

Garber, K.B., Visootsak, J. & Warren, S.T. (2008). Fragile X-syndrome. *European Journal of Human Genetics, 16,* 666-672.

Gillberg, C.L. (1992). Autism and autistic-like conditions: subclasses among disorders of empathy. *Journal of Child Psychology and Psychiatry, 33,* 813-842.

Gillberg, C. & Billstedt, E. (2000). Autism and Asperger syndrome: coexistence with other clinical disorders. *Acta Psychiatrica Scandinavia, 102,* 321-330.

Gillberg, I.C. & Gillberg, C. (1989). Asperger syndrome-some epidemiological considerations: a research note. *Journal of Child Psychology and Psychiatry, 30,* 631-638.

Gonzales, M.L. & LaSalle, J.M. (2010). The role of MeCP2 in brain development and neurodevelopmental disorders. *Current Psychiatry Reports, 12,* 127-134.

Guilmatre, A., Dubourg C., Mosca, A.L., Legallic S., Goldenberg, A., Drouin-Garraud, V., et al. (2009). Recurrent rearrangements in synaptic and neurodevelopment genes and shared biologic pathways in schizophrenia, autism, and mental retardation. *Archives of General Psychiatry, 66,* 947-956.

Green, J.J. & Hollander E. (2010). Autism and ocytocin: new developments in translational approaches to therapeutics. *Neurotherapeutics, 7,* 250-257.

Gregory, S.G., Connelly, J.J., Towers, A.J., Johnson, J., Biscocho, D., Markunas, C.A., et al. Genomic and epigenetic evidence for oxytocin receptor deficiency in autism. *BMC Medicine, 7,* 62.

Hagerman, R.J. (2005). Fragile X syndrome. In S.B. Cassidy & J.E. Allanson (eds.) *Management of Genetic Syndromes* (pp 251-263). Hoboken, NJ: Wiley-Liss.

Happé, F. & Frith, U. (2006). The weak coherence account: Detail-focused cognitive style in Autism Spectrum Disorders. *Journal of Autism and Developmental Disorders, 36,* 5-25.

Hernandez, R.N., Feinberg, R.L., Vaurio, R., Passanante N.M., Thompson, R.E. & Kaufmann W.E. (2009). Autism Spectrum Disorder in Fragile X syndrome: a longitudinal evaluation. *American Journal of Medical Genetics, 149A,* 1125-1137.

Hessl, D., Rivera, S.M. & Reiss, A.L. (2004). The neuroanatomy and neuroendocrinology of fragile X syndrome. *Mental Retardation and Developmental Disabilities Research Reviews, 10,* 17-24.

Hogart, A., Wu, D., LaSalle, J.M. & Schanen, M.C. (2010). The comorbidity of autism with the genomic disorders of chromosome 15q11.2-q13. *Neurobiology of Disease, 38,* 181-191.

Hollander, E., Bartz, J., Chaplin, W., Philips, A., Summer, J., Soorya, L., et al. (2007). Oxytocin increases retention of social cognition in autism. *Biological Psychiatry, 61,* 498-503.

Insel, T.R. (2010). The challenge of translation in social neuroscience: a review of oxytocin, vasopressin and affiliative behavior. *Neuron, 65,* 768-779.

Jacquemont, M.L., Sanlaville, D., Redon, R., Raoul, O., Cormier-Daire, V., Lyonnet, S., et al. (2006). Array-bases comparative genomic hybridisation identifies high frequency of cryptic chromosomal rearrangements in patients with syndromic autisme spectrum disorders. *Journal of Medical Genetics, 43,* 843-849.

Jamain, S., Quach, H., Betancur, C., Rástam, M., Colineaux, C., Gillberg, I.C., et al. (2003). Mutations of the X-linked genes encoding neuroligins NLGN3 and NLGN4 are associated with autism. *Nature Genetics, 34,* 27-29.

Kanner, L. (1971). Follow-up study of eleven autistic children originally reported in 1943. *Journal of Autism and Childhood Schizophrenia, 1,* 199-145.

Kanner, L. (1943). Autistic disturbances of affective contact. *The Nervous Child, 2,* 17-50.

Khetrapal, N. (2008). The framework for disturbed affective consciousness in autism. *Neuropsychiatric Disease and Treatment, 4,* 531-533.

Kretschmer, E. (1921) *Körperbau und Charakter*. Berlin: Springer.

Kumar, R.A. & Christian, S.L. (2009). Genetics of autism spectrum disorders. *Current Neurology and Neuroscience Reports, 9,* 188-197.

Kumbier, E., Domes, G., Herpertz-Dahlmann, B. & Herpertz, S.C. (2010). Autismus und autistische Störungen. *Der Nervenarzt, 81,* 55-65.

Lee, H.J., Macbeth, A.H., Pagani, J.H. & Scott Young, W. (2009). Oxytocin: The great facilitator of life. *Progress in Neurobiology, 88,* 127-151.

Leonhard, K. (1957) *Aufteilung der endogenen Psychosen,* Berlin: Akademie-Verlag.

Lerer, E., Levi, S., Salomon, S., Darvasi, A., Yirmiya, N. & Ebstein, R.P. (2008). Association between the oxytocin receptor (OXTR) gene and autism: relationship to Vineland Adaptive Behavior Scales and cognition. *Molecular Psychiatry, 13,* 980-988.

Lichtenstein, P., Carlström, E., Rastam, M., Gillberg, C., Anckersäter, H. (2010) The gentics of autism spectrum disorders and related neuropsychiatric disorders in childhood. *American Journal of Psychiatry, 167,* 1357-1363.

Lintas, C. & Persico, A.M. (2009). Autistic phenotypes and genetic testing: state-of-the-art for the clinical geneticist. *Journal of Medical Genetics, 46,* 1-8.

Lisé, M.F. & El-Husseini, A. (2006). The neuroligin and neurexin families: from structure to function at the synapse. *Cellular and Molecular Life Sciences, 63,* 1833-1849.

Losh, M., Sullivan, P.F., Trembath D. & Piven, J. (2008). Current developments in the genetics of autism: from phenome to genome. *Journal of Neuropathology and Experimental Neurology, 67,* 829-837.

Lubs, H.A. (1969) A marker X chromosome. *American Journal of Human Genetics, 21,* 231-244

Manzi, B., Loizzo, A.L., Giana, G. & Curatolo, P. (2008). Autism and metabolic diseases. *Journal of Child Neurology, 23,* 307-314.

Marco E.L. & Skuse, D.H. (2006). Autism-lessons from the X chromosome. *Scan, 1,* 183-193.

Matijevic, T., Knezevic, J., Slavica, M. & Pavelic, J. (2009). Rett Syndrome: from the gene to the disease. *European Neurology, 61,* 3-10.

Matuszek, G & Talebizadeh, Z. (2009). Autism genetic database (AGD): a comprehensive database including autism susceptibility gene-CNV's integrated with known noncoding RNAs and fragile sites. *BMC Medical Genetics, 10,* 102.

Marcotte, L. & Crino, P.B. (2006). The neurobiology of the tuberous sclerosis complex. *Neuromolecular Medicine, 8,* 531-546.

McHugh, L., Barnes-Holmes, F., & Barnes, D. (2004). A Relational Frame Account of the Development of Complex Cognitive Phenomena: Perspective-taking, False Belief Understanding, and Deception. *International Journal of Psychology and Psychological Therapy, 4,* 303-324.

Moss, L & Howlin, P. (2009). Autism spectrum disorders in genetic syndromes: implications for diagnosis, intervention and understanding the wider autism spectrum disorder population. *Journal of Intellectual Disability Research, 53,* 852-873.

Nasrallah, F., Feki, M., & Kaabachi N. (2010). Creatine and creatine deficiency syndromes: biochemical and clinical aspects. *Pediatric Neurology, 42,* 163-171.

Niklasson, L., Rasmussen, P., Oskarsdóttir, S. & Gillberg, C. (2009). Autism, ADHD, mental retardation and behavior problems in 100 individuals with 22q11 deletion syndrome. *Research in Developmental Disabilities, 30,* 763-773.

Orlova, K.S. & Crino, P.B. (2010). The tubereous sclerosis complex. *Annals of the New York Academy of Science, 1184,* 87-105.

Paylor, R., Glaser B., Mupo A., Ataliotis P., Spencer C., Sobotka, A. et al. (2006). Tbx1 haploinsufficiency is linked to behavioral disorders in mice and humans: implications for 22q11 deletion syndrome. *Proceedings of the National Academy of Sciences, 103,* 7729-7734.

Piggot, J., Shirinyan D., Shemmassian, S., Vazarian, S & Alarcon, M. (2009). Neural systems approaches to the neurogenetics of autism spectrum disorders. *Neuroscience, 164,* 247-256.

Rett A. (1966) Uber ein cerebral-atrophisches syndrom bei hyperammonämie. *Wiener Medischer Wochenschrift, 116,* 723-726.

Rice, C. (2009). Prevalence of autism spectrum disorders - Autism and developmental disabilities monitoring network, United States, 2006. *MMWR Surveillance Summaries, 58,* 1-20.

Rosenberg, E.H., Almeida L.S., Kleefstra, T., deGrauw, R.S., Yntema, H.G., Bahi, N., et al., (2004). High prevalence of SLC6A8 deficiency in X-linked mental retardation. *American Journal of Human Genetics, 75,* 97-105.

Rutter, M. (1968). Concepts of autism: a review of research. *Journal of Child Psychology and Psychiatry, 9,* 1-25.

Salomons, G.S., van Dooren S.J.M., Verhoeven, N.M., Cecil, K.M., Ball, W.S. Degrauw T.J., et al. (2001). X-linked creatine-transporter gene (SLC6A8) defect: a new creatine-deficiency syndrome. *American Journal of Human Genetics, 68,* 1497-1500.

Salomons, G.S., van Dooren, S.J.M., Verhoeven, N.M., Marsden, D., Schwartz, C., Cecil, K.M., et al. (2003). X-linked creatine transporter defect: an overview. *Journal of Inherited Metabolic Disorders, 26,* 309-318.

Sebat, J., Lakshmi, B., Malhotra, D., Troge, J., Lese-Martin, C., Walsh, T., et al. (2007). Strong association of de novo copy number mutations with autism. *Science, 316,* 445-449.

Shprintzen, R.J., Goldberg R.B., Lewin, M.L., Sidoti, E.J., Berkman, M.D., Argamaso, R.V., et al. (1978) A new syndrome involving cleft palate, cardiac anomalies, typical facies, and learning diabilities: velo-cardio-facial syndrome. *Cleft Palate, 15,* 56-62.

Shprintzen, R.J. (2000). Velo-cardio-facial syndrome: a distinctive behavioural phenotype. *Mental Retardation and Developmental Disabilities, 6,* 142-147.

Szatmari, P, Paterson, A.D., Zwaigenbaum, L., Roberts, W., Brian, J., Liu, X.Q., et al. (2007). Mapping autism risk loci using genetic linkage and chromosomal rearrangements. *Nature Genetics, 39,* 319-328.

Veltman, J.A. & De Vries, B.A. (2006). Unbased whole genome or targeted analysis. *Journal of Molecular Diagnostics, 8,* 534-537.

Van Amelsvoort, T., Daley, E., Robertson, D., Ng, V., Owen, M., Murphy, K.C., et al. (2004). Brain anatomy in adults with velocardiofacial syndrome with and without schizophrenia. *Archives of General Psychiatry 61,* 1085-1096.

Verhoeven, W.M.A., Tuinier, S. & van der Burgt, I. (2008). Top-down or bottom-up: contrasting perspectives on psychiatric diagnoses. *Biologics: Targets & Therapy, 2,* 409-417.

Verhoeven, W. Egger, J. & Tuinier, S. (2007). Thoughts on the behavioural phenotypes in Prader-Willi syndrome and velo-cardio-facial syndrome: a novel approach. *Acta Neuropsychiatrica, 19,* 244-250.

Verhoeven, W.M.A., Csepan, R., Marcelis, C., Lafeber, D., Egger, J.I.M., & Tuinier, S. (2010). Sanfilippo B in an elderly female psychiatric patient: a rare but relevant diagnosis in presenile dementia. *Acta Psychiatrica Scandinavica, 122,* 162-165.

Vissers, L.E., de Ligt, J., Gilissen, C., Janssen, I., Steehouwer, M., et al. (2010). A de novo paradigm for mental retardation. *Nature Genetics,* 42, 1109-1112.

Vogels, A., Verhoeven, W.M.A., Tuinier, S., DeVriendt, K., Swillen, A., Curfs, L.M.G., et al. (2002). The psychopathological phenotype of velo-cardio-facial syndrome. *Annales de Génétique, 45,* 89-95.

Volkmar, F.R., State, M. & Klin, A. (2009). Autism and autism spectrum disorders: diagnostic issues for the coming decade. *Journal of Child Psychology and Psychiatry, 50,* 108-115.

Vorstman, J.A.S., Staal W.G., van Daalen, E., van Engeland, H., Hochstenbach, P.F.R. & Franke, L. (2006a). Identification of novel autism candidate regions through analysis of reported cytogenetic abnormalities associated with autism. *Molecular Psychiatry, 11,* 18-28.

Vorstman, J.A.S., Morcus, M.E.J., Duijff, S.N., Klaassen, P.W.J., Heineman-de Boer, J.A., Beemer, F.A. et al. (2006b). The 22q11.2 deletion in children: high rate of autistic disorders and early onset of psychotic symptoms. *Journal of American Academy of Child and Adolescent Psychiatry, 45,* 1104-1113.

Weiss, L.A., Arking, D.E., The Gene discovery Project of Johns Hopkins & The Autism Consortium (2009). A genome-wide linkage and association scan reveals novel loci for autism. *Nature, 461,* 802-808.

Weiss, L.A., Shen, Y., Korn, J.M., Arking, D.E., Miller, D.T., Fossdal, R., et al. (2008). Association between microdeletion and microduplication at 16p11.2 and autism. *The New England Journal of Medicine, 358,* 667-675.

Weissman, J.R., Kelley, R.I., Bauman, M.L., Cohen, B.H., Murray, K.F., Mitchell, R.L., et al. (2008). Mitochoncrial disease in autism spectrum disorder patients: a cohort analysis. *Plos One,* 3, e3815.

Wing, L. (1981). Asperger's syndrome: a clinical account. *Psychological Medicine, 11,* 115-129.

Wing, L. (2005). Reflections on opening Pandora's box. *Journal of Autism and Developmental Disorders, 35,* 197-203.

Wing, L. & Gould, J. (1979). Severe impairments of social interaction and associated abnormalities in children: epidemiology and classification. *Journal of Autism and Developmental Disorders, 9,* 11-29.

Wiznitzer, M. (2004). Autism and tuberous sclerosis. *Journal of Child Neurology, 19,* 675-679.

Zecavati, N. & Spence, S.J. (2009). Neurometabolic disorders and dysfunction in autism spectrum disorders. *Current Neurology and Neuroscience Reports, 9,* 129-136.

Part 3

Genetics and Pathophysiology of Autism Spectrum Disorders

4

The Genetics of Autism Spectrum Disorders

John J.M. Connolly[1] and Hakon Hakonarson[1,2]
[1]*Center for Applied Genomics*
The Children's Hospital of Philadelphia, Philadelphia
[2]*Department of Pediatrics,University of Pennsylvania School of Medicine, Philadelphia*
USA

1. Introduction

Autism is a neurodevelopmental disorder of complex etiology and is amongst the most heritable of neuropsychiatric disorders while sharing genetic liability with other neurodevelopmental disorders such as intellectual disability (ID). Autism spectrum disorders (ASDs) are defined more broadly and include autism, Asperger syndrome, childhood disintegrative disorder and pervasive developmental disorder not otherwise specified. Under the Diagnostic and Statistical Manual of Mental Disorders, 4th Edition Revised (DSM-IVTR), these disorders are grouped together with Rett syndrome ("Rett's disorder") as pervasive developmental disorders. However, Rett syndrome has a reportedly distinct pathophysiology, clinical course, and diagnostic strategy (Levy & Schultz, 2009) and will likely be removed in the impending publication of DSM-V (APA, 2010). The new diagnostic manual will formally adopt the single diagnostic category "ASDs", which is used here. Reported prevalence rates for ASDs range from 20 (Newschaffer *et al.* 2007) to 116 (Baird *et al.*, 2006) per 10,000 children, and vary in accordance with diagnostic, sampling, and screening criteria. The Centers for Disease Control and Prevention (CDC) suggest that in the United States, the prevalence of ASDs is 1 in 110 (1/70 in boys and 1/315 in girls) (ADDM, 2009). The three primary characteristics of ASDs are communication impairments, social impairments, and repetitive/stereotyped behaviors. The DSM-IVTR, ICD-10, and many other diagnostic instruments require impairment in each of these domains for a diagnosis of autistic disorder.

Within the last decade, a number of major technological developments have transformed our understanding of the genetic causes of autism, and the field continues to evolve rapidly. In this chapter, we will review three approaches to identifying genetic factors that contribute to the pathogenesis of ASDs: 1) common variants and genome-wide association studies (GWAS); 2) rare variants and copy number variation (CNV) studies, and 3) familial forms of autism and the role of next-generation sequencing (NGS) methods. Data from all three approaches underscores the conclusion that autism is a highly complex and heterogeneous disorder, involving a multifactorial etiology. Moreover, it is becoming increasingly apparent that autism is not a unitary disorder, and that the spectrum may consist of any number of different autisms that share similar symptoms or phenotypes. This conclusion has important implications for evaluation and treatment, which are discussed in the conclusion.

Genome-wide association studies (GWAS) examine the frequency of single nucleotide polymorphisms (SNPs) in cases *versus* control populations and can adopt either a case-control or family-based approach. The former allows researchers to avoid the often complex process of acquiring diagnostic/phenotype data from a patient's family, and can incorporate very large numbers of control datasets that may be more readily available. The latter controls for the often confounding phenomenon of population stratification, where variants more common to specific racial groups may either be erroneously identified as causal, or obscure actual causal variants. A major caveat with family-based designs is the often unfounded assumption that unaffected family members do not share causal variants.

GWAS test for common variants (>1% population frequency), with the assumption that ASDs are at least in part caused by the coinheritance of multiple risk variants, each of small individual effect (odds ratios between 1:1 and 1:5). This assumption is known as the common disease-common variant (CDCV) model (Risch and Merikangas, 1996).

A 2009 paper by Wang *et al.* (2009) from our laboratory was the first to identify common variants for ASDs on a genome-wide scale. We examined 780 families (3,101 individuals) with affected children, a second group of 1,204 affected individuals, and 6,491 controls, all of whom were of European ancestry. We identified six genetic markers on chromosome 5 in the 5p14.1 region that confirmed susceptibility to ASDs. The region straddles two genes, *CDH9* and *CDH10*. Both genes encode type II classical cadherins, transmembrane proteins that promote cell adhesion. The association of cadherins is consistent with the cortical-disconnectivity model of autism (e.g. Gepner & Féron, 2009), which postulates that ASDs may result from an increase or decrease in functional connectivity and neuronal synchronization in relevant neural pathways. Functional studies suggest that under-activity between and within networks are correlated with social, communication, cognitive, and sensorimotor impairments (Müller *et al.*, 2011).

The study design utilized two independent replication cohorts and the key SNP at this locus has also been replicated in two additional independent cohort studies (Ma *et al.*, 2009; St Pourcain *et al.*, 2010), lending further support that genetic factors at the 5p14 locus, which is flanked by two relevant cadherin genes, represent strong candidates for aligning molecular function with known neural deficits in ASDS. The original report by Wang and colleagues, also demonstrated that there was an enrichment in Catherin associated genes in ASDs in general, based on gene-pathway analysis (Wang *et al*, 2009). Cadherins represent a large family of transmembrane proteins that mediates calcium-dependent cell–cell adhesion and, *via* cell adhesion, has been shown to generate synaptic complexity in the developing brain (Redies, 2000). Other common GWAS variants reported have not been replicated in independent studies and will not be covered here.

4.1 Replicated common variants from candidate gene studies

Other common variants from candidate gene studies include CNTNAP2, EN2, and MET, which are reviewed briefly below. A more in depth review of these genes can be derived from catalogs at http://www.genome.gov/26525384 and http://w w w.ncb.nlm.nih.gov/o mim/209850.

Located on chromosome 7q35, Contactin Associated Protein 2 (*CNTNAP2*) was identified by Alarcón *et al.* (2002) as a candidate for the age at first word endophenotype. A subsequent follow-up by the same group (Alarcón *et al.* 2008) using linkage, association, and gene-expression analyses, found *CNTNAP2* to be the only autism-susceptibility gene to reach significance across all approaches. An independent linkage analysis by Arking *et al.* (2008) also highlighted *CNTNAP2* as a significant ASD candidate gene. *CNTNAP2* is part of the neurexin family, which have repeatedly been associated with autism (see below). Interestingly, Vernes *et al.* (2008) showed that *CNTNAP2* binds to *FOXP2*, which is a well-established correlate of language and speech disorders (Lai *et al.*, 2001) – a common phenotype in ASDs.

Engrailed 2 (*EN2*) is a homeobox gene that is critical to the development of the midbrain and cerebellum. Like other homeobox genes, it regulates morphogenesis. *EN2* is a human homolog of the engrailed gene, which is found in Drosophila. *En2* mouse mutants have anatomic phenotypes in the cerebellum that resemble cerebellar abnormalities reported in autistic individuals (Cheng *et al.*, 2010). Benayed *et al.* (2005, 2009) have reported and replicated in three separate datasets a significant association with broad and narrow ASD phenotypes. Wang *et al.* (2008) also found an association between *EN2* and ASDs in a Chinese Han sample, although Zhong *et al.* (2003) failed to find evidence of an underlying association.

The oncogene *MET* is also strongly linked to ASD etiology, having been supported by a number of studies in the past decade (e.g. IMGSAC, 2001; Campbell *et al.*, 2006, 2008; Sousa *et al.*, 2009). Recently, Eagleson *et al.* (2011) reported a role for *Met* signaling in cortical interneuron development in vitro in a mouse model.

4.2 Unexplained variance

For the most significant discovery SNP identified in the Wang *et al.* study above (rs4307059), the risk allele frequency was 0.65 in cases with an odds ratio of 1.19, which is comparable with common variant discoveries in other psychiatric disorders including schizophrenia (Glessner & Hakonarson, 2009; Glessner *et al.*, 2010); bipolar disorder (Ferreira, 2008), and attention-deficit/hyperactivity disorder (Arcos-Burgos *et al.* 2010). While it is important not to undermine the significance of these findings, it should be noted that the predictive value of such ratios is relatively low (Dickson *et al.* 2010), often explaining less than 5% of the total risk (review at http://www.genome.gov/26525384). However, it is also possible that these common SNPs may be tagging a more rare causative variant (i.e., synthetic association), where the effect sizes may be markedly underestimated by the GWAS variant as we recently reported (Dickson *et al.* 2010). In one example, Wang *et al.* (2010), examined the *NOD2* locus as a cause of Crohn disease. Using resequencing data, they found that three causal variants explain > 5% of the genetic risk, where GWAS had estimated the risk at ~1%. Careful phenotyping of cohorts is important to ensure that the phenotypes produced by rare-variants are not being "filtered-out"and thereby missed as a consequence. A long range haplotype analysis of the GWAS data at the respective loci is therefore recommended in an attempt to enrich for individuals with rare-causative variants, who should be picked out of the cohort and subsequently sequenced for confirmation (Wang *et al.*, 2010). This is clearly a critical point, particularly in relation to psychiatric disorders, where diagnoses can be more contingent upon subjective observation than, for example, the genetics of height, which can utilize more intrinsically quantitative data.

The possibility that common variants are not the major cause of ASDs is also gaining increased support from the preponderance of copy number variation (CNV) studies, which are identifying rare variants with a stronger causal impact.

5. Copy number variation in ASDs

Copy number variations (CNVs) are insertions, deletions, or translocations in the human genome that are universal in the general population but more commonly found in genic regions in individuals with neuropsychiatric disorders (e.g. Pinto *et al.*, 2010). CNVs can be detected by the same SNP arrays used in GWAS, and vary in length from many megabases to 1 kilobase or smaller. They are often not associated with any observable phenotype.

One of the most widely-known CNVs is Down syndrome, which is characterized by an extra chromosome 21. Rett syndrome is also caused by a CNV, which includes a deletion in *MECP2*. CNVs can be inherited or occur *de novo*, the cause of which is thus far unknown. Common disease-causing CNVs are infrequent but rare CNVs, with a frequency of less than 1%, have been identified for a range of disorders including ADHD (e.g. Williams *et al.*, 2010), schizophrenia (e.g. Glessner *et al.*, 2010; Levinson *et al.*, 2011), bipolar disorder (e.g. Chen *et al.*, 2010) and many others. A substantial portion of autism appears to be caused by rare CNVs. *De novo* CNVs that are greater than 100kb in size are more common in genic regions in individuals with ASDs than in the general population.

Sebat *et al.* (2007) provided some relevant early insights into the genomic features of CNVs. Firstly, they noted that *de novo* CNVs were individually rare - from 118 ASD cases, none of the identified variants were observed more than twice, with the majority seen just once. This confirmed the widely-held assumption that many different loci can contribute to the same ASD phenotype. Secondly, the authors affirmed the utility of population-study approaches that analyze sporadic and multiplex (i.e. more than one family member affected) families separately. The rate of *de novo* mutation in large (mostly genic) loci in multiplex families was significantly lower than for the sporadic cases ($p = 0.04$). While this observation remains to be replicated in a larger study, the finding implies two mechanisms of genetic susceptibility - spontaneous mutation and inheritance. Finally, the sheer volume of loci identified by this approach (multiple loci on 20 chromosomes) affirms the extraordinarily complexity of ASDs.

A number of subsequent studies have greatly expanded the number of candidate loci. Our laboratory (Bucan *et al.* (2009)) reported 150+ CNVs in 912 ASD families that were not found in 1,488 controls. Critically, 27 of these loci were replicated in an independent cohort of 859 ASD cases and 1,051 controls. Some of the rare variants we identified had previously been associated with autism, including *NRXN1* and *UBE3A*, which are established ASD candidate genes (Guilmatre *et al.*, 2009). Samaco *et al.* (2005) previously identified significant deficits in *ube3a* expression in *mecp2*-deficient mice, suggesting a shared pathological pathway with Rett syndrome (as well as Angelman syndrome, and autism). Similarly, Kim *et al.* (2008) associated *NRXN1* with a balanced chromosomal abnormality at chromosome 2p16.3 in two unrelated ASD individuals. Rare variants in the coding region included two missense changes.

Glessner *et al.* (2009) identified and reported CNVs in two major gene networks, including neuronal cell adhesion molecules (such as *NRXN1*) and the ubiquitin gene family (such as *UBE3A*). Interestingly, four of the most prominent genes enriched by CNVs in ASD cases (*UBE3A*, *PARK2*, *RFWD2* and *FBXO40*) are all part of the ubiquitin gene family. Ubiquitination can alter protein function after translation, and degrade target proteins in conjunction with proteasomes. The ubiquitin–proteasome system operates at pre- and post-synapses, whose functions includes regulating neurotransmitter release, recycling synaptic vesicles in pre-synaptic terminals, and modulating changes in dendritic spines and post-synaptic density (Yi & Ehlers, 2005). As well as implicating an ubiquitination network in relation to ASDs, we also identified a second pathway involving *NRXN1*, *CNTN4*, *NLGN1*, and *ASTN2*. Genes in this group mediate neuronal cell-adhesion, and contribute to neurodevelopment by facilitating axon guidance, synapse formation and plasticity, and neuron–glial interactions. We also note that ubiquitins are involved in recycling cell-adhesion molecules, which is a possible mechanism by which these two networks are cross linked.

In a similar approach, Pinto *et al.* (2010) further confirmed the importance of rare CNVs as causal factors for ASDs. Interestingly, the group did not observe a significant difference between cases and controls in terms of raw number of CNVs or estimated CNV size. However, the number of CNVs in genic regions was significantly greater in ASDs compared to controls. Again, loci enriched for CNVs include a number of genes known to be important for neurodevelopment and synaptic plasticity, such as *SHANK2*, *SYNGAP1*, and *DLGAP2*. Between 5.5% and 5.7% of ASD cases have at least one *de novo* CNV, further confirming the significance of *de novo* genetic events as risk factors for autism. Similar to the Glessner study, the Pinto group mapped CNVs to a series of networks involved in the development and regulation of the central nervous system functions. Implicated networks include neuronal cell adhesion, GTPase regulation (important for signal transduction and biosynthesis), and GTPase/Ras signaling, also involved in ubiquitination.

Finally, Gai *et al.* (2011) took a slightly different approach, focusing exclusively on inherited CNVs. While underlying loci were not necessarily common to those identified by the Glessner and Pinto groups, enrichment in pathways involving central nervous system development, synaptic functions and neuronal signaling processes was again confirmed. The Gai *et al.* study also emphasized the role of glutamate-mediated neuronal signals in ASDs.

Collectively, these CNV studies suggest that certain hotspots on the genome are particularly vulnerable to ASDs, which include loci on chromosomes 1q21, 3p26, 15q11-q13, 16p11, and 22q11. These hotspots are part of large gene networks that are important to neural signaling and neurodevelopment and have additionally been associated with other neuropsychiatric disorders.

In particular, a number of CNV studies in schizophrenia have highlighted structural mutations incorporating chromosomes 1q21, 15q13, and 22q11 (e.g. McClellan and King 2010; Glessner *et al.*, 2010), which are significantly enriched in cases versus controls, with *NRXN1* being a standout in this regard. From a phenotype perspective, autism and schizophrenia seem very different, both in behavioral manifestation and age of onset, and it may seem counter-intuitive that associated loci should overlap. Some authors have addressed this peculiarity by proposing that schizophrenia and autism may in fact be different poles of the same spectrum. Thus, Crespi and Braddock (2008) suggest that social cognition is underdeveloped in ASDs and over-developed in the psychotic spectrum, with a similar polarization of language and behavioral phenotypes. Although speculative, this hypothesis has gained some traction. In the next several years, genomic, imaging, and model-systems approaches will likely shed further light on the relationship between autism, schizophrenia and other neuropsychiatric disorders.

6. Sequencing familial forms of ASDs

To this point, we have focused primarily on the complex interactions of polygenic networks as the major cause of ASDs. However, this is not exclusively the case. Paralleling the recent spate of CNV is a renewed focus on rare disorders, including familial forms of complex diseases that potentially are monogenic or with less complex inheritance pattern. At the outset of this chapter, we emphasized the overlap with fragile X syndrome, where one third of cases are co-morbid for ASD. As mentioned, fragile X is caused by a failure to express the

protein coded by *FMR1*. However, mutations in *FMR1* do not always result in fragile-X and can result in a phenotype more representative of ASDs. Thus, Muhle *et al.* (2004) found that 7-8% of idiopathic ASD cases may have mutations at the *FMR1* locus. Likewise, although mutations in *MECP2* are the common cause of Rett syndrome, certain mutations at the same locus have been associated with idiopathic autism (Carney *et al.* (2003).

X-linked genes encoding neurologins *NLGN3* and *NLGN4* and *SHANK3* (a neuroligin binding partner) are other prominent examples of distinct rare genetic causes, and a parallel can be drawn with these studies and mental retardation and epilepsy, which include many rare syndromes that collectively account for a substantial proportion of the two disorders (Morrow *et al.*, 2008). Indeed it is perhaps more than coincidence that autism is heavily co-morbid with these two conditions, with >40%(Bölte *et al.*, 2009) and ~40% (Danielsson *et al.*, 2005) of ASD cases meeting diagnostic criteria for mental retardation and epilepsy respectively. It also is noteworthy that many of these monogenic-related genes are also major players in neurodevelopment and synapse activity. Other prominent examples include *TSC1*, *TSC2* (Osborne *et al.*, 1991; Franz, 1998), *NF1*, and *UBE3A* (see Morrow *et al.* (2008).

The identification of monogenic or possibly oligogenic autisms is likely to accelerate in the next several years as next-generation sequencing becomes more widely available. We recently encountered a family of two parents, six healthy siblings, and two siblings with severe autism suggestive of autosomal recessive inheritance. Unsuccessful attempts using linkage and CNV approaches failed to identify a causal locus, but whole-exome sequencing at 20x coverage identified four genes, including one with a non-synonymous SNP in the protocadherin alpha 4 isoform1 precursor (*PCDHA4*) gene, which presents a strong candidate gene, currently under validation. Protocadherins are part of the cadherin family that facilitates neuronal cell adhesion and this discovery is consistent genomically and neurobiologically with the findings addressed above in relation to *CDH9* and *CDH10*.

Known syndromes with ASD features include fragile-x, neurofibromatosis type 1, down syndrome, tuberous sclerosis, neurofibromatosis (which confers a 100-fold increased risk for ASDs Li *et al.* (2005), Angelman, Prader-Willi and related 15q syndromes, and at least several dozen others (see Zafeiriou *et al.*, 2007 for a comprehensive review). **Table 1** from Volkmar *et al.* (2005) lists the most commonly associated syndromes with median rate and range. It is likely that many more unidentified rare syndromes with Mendelian causes have ASD phenotypes. As of March 2011, the Online Mendelian Inheritance in Man (OMIM) database listed 6,727 known or suspected Mendelian diseases (MD), with 2,993 (44%) of these having an identified molecular basis. Since OMIM derives its data from published reports, these figures likely under-represent rare disorders, which may go unreported. It has been proposed that as many as 30,000 genetic disorders may exist, suggesting that many Mendelian disorders have no genetic etiology identified to date. Given the large-representation of autism phenotypes in known syndromes, we can assume a similar trend in unreported disease.

It remains to be determined whether rare variants will account for the majority of autisms. Irrespective, as with many other aspects of scientific inquiry, the study of rare variants will continue to play an important role in explicating the pathogenesis of ASDs. El-Fishawy and State (2010) point to hypercholesterolemia and hypertension (Brown, 1974; Lifton *et al.*, 2001) as examples where rare mutations have been successful in driving a molecular understanding of the disease as opposed to identifying risk factors in the general population. Rare mutations, particularly when they are Mendelian, carry large effects and are typically in genic regions. These characteristics make the resolution of underlying networks distinctly less complex and, moreover, are amenable to modeling in other systems. Recent groundbreaking studies by Marchetto *et al.* (2010) and Muotri *et al.* (2010), who created a cell culture model of Rett syndrome, are potentially exciting developments in this regard. Here, the researchers used skin biopsies from four Rett's patients, each carrying a different *MeCP2* mutation, to culture induced pluripotent stem cells (iPS). Once the iPS cells developed into neurons, they showed a decreased number of neurons and dendritic spines, consistent with neurodevelopmental disruptions. Intervention with insulin-like growth factor 1 (*IGF1*), which is known to regulate neurodevelopment, was subsequently shown to reverse Rett-like symptoms in a mouse model of the disease. This innovative approach is an exciting model of how rare gene approaches can stimulate our understanding of the pathophysiology and potential reversibility of ASDs.

Syndrome	Number of Studies	Median Rate	Range %
Tuberous sclerosis	11	1.1	0–3.8
Fragile X	9	0.0	0–8.1
Down syndrome	12	0.7	0–16.7
Neurofibromatosis 1	6	0	0–1.4

Table 1. Associated disorders and their rate in autism (from Volkmar *et al.*, 2005 in Zafeiriou *et al.* 2007)

7. Conclusions

ASDs are clearly highly heritable disorders and advances in gene-finding technology in the past decade have rapidly accelerated gene discovery. As is typically the case, successive developments have made the problem more complex such that there are dozens of candidate genes, many of which remain to be replicated. In spite of this complexity, we can observe a number of patterns beginning to unfold 1) the relative scarcity of causal common variants, 2) the growing list of causal rare variants, and 3) the emergence of monogenic disorders with primary and secondary ASD phenotypes.

The monogenic autisms are particularly interesting from a treatment perspective, as they provide a mechanism for studying ASD phenotypes in model systems and an obvious target for drug intervention. They are also amenable to clinical testing and the decreasing cost of research technologies means that this capacity is more widely available to clinicians. In fact, as the resolution of clinical instruments becomes more sophisticated, it is likely that the clinic will become a primary workplace for syndromic discovery.

A key requirement in driving gene discovery is the necessity of high-quality phenotype data. ASDs are notoriously heterogeneous, and are fractionated in terms of symptoms and trajectory. Mandy & Skuse (2008) reviewed seven factor analysis studies of ASDs symptoms, and found that all but one dissociated social and non-social factors. In a non-clinical sample of 3,000 twin pairs, Happé *et al.* (2006) examined autistic-like traits and found consistently low correlations ($r = 0.1$-0.4) between each of the core deficits on the autism spectrum. Endophenotypes, sub-components or sub-processes of the broader phenotype, may provide a productive avenue to disentangling some of this complexity. By filtering out all but a few discrete measures, we can theoretically increase the signal-to-noise ratio in genotype-phenotype associations. A number of endophenotypes for ASDs have been identified associated with disease genes, including head circumference (associated with the *HOXA1* A218G polymorphism, Conciatori *et al.*, 2004), age at first word (associated with a quantitative trait locus on 7q35, Alarcón *et al.* 2005), delayed magnetoencephalography evoked responses to auditory stimuli (Roberts *et al.*, 2010), and enhanced perception (Mottron *et al.*, 2006). The endophenotype approach is arguably more consistent with rare-/mono-genic discovery, where a mutated network may not yield a diagnosis of autism *per se*, but nevertheless cause associated abnormalities. Note, this approach does not diminish the pleiotropic effects of genes involved in neurodevelopment, and only serves to make the point that the relevant genotype may associate with some but not all ASD features.

The converse, of course, is also true with a large number of candidate genes contributing to the majority of known ASDs. With ~80% of genes expressed in the brain it is likely that this number will continue to grow, and here again careful phenotyping is critical to identifying functional consequences. Ultimately, the primary goal is not to determine the frequency of variation/mutation in cases *versus* controls, but to determine the pathway(s) and gene networks that lead to pathology. This will be no mean feat, with other major players such as epigenetic factors, RNA regulatory elements, and environmental exposures also an important part of the equation. While daunting, the elucidation of these elements will doubtlessly take us closer to developing effective treatments for ASDs. Given the current rate of progress, we have cause for cautious optimism in this regard.

8. References

Alarcon, M., Abrahams, B. S., Stone, J. L., Duvall, J. A., Perederiy, J. V., Bomar, J. M., Geschwind, D. H. (2008). Linkage, association, and gene-expression analyses identify CNTNAP2 as an autism-susceptibility gene. *Am J Hum Genet, 82*(1), 150-159. doi: 10.1016/j.ajhg.2007.09.005

Alarcon, M., Yonan, A. L., Gilliam, T. C., Cantor, R. M., & Geschwind, D. H. (2005). Quantitative genome scan and Ordered-Subsets Analysis of autism endophenotypes support language QTLs. *Mol Psychiatry, 10*(8), 747-757. doi: 10.1038/sj.mp.4001666

Alarcon, M., Cantor, R. M., Liu, J., Gilliam, T. C., & Geschwind, D. H. (2002). Evidence for a language quantitative trait locus on chromosome 7q in multiplex autism families. *Am J Hum Genet, 70*(1), 60-71. doi: 10.1086/338241

Arcos-Burgos, M., Jain, M., Acosta, M. T., Shively, S., Stanescu, H., Wallis, D., Muenke, M. (2010). A common variant of the latrophilin 3 gene, LPHN3, confers susceptibility to ADHD and predicts effectiveness of stimulant medication. *Mol Psychiatry, 15*(11), 1053-1066. doi: 10.1038/mp.2010.6

American Psychiatric Association, (2010, March 22, 2010). New Proposed Changes Posted for Leading Manual of Mental Disorders. Retrieved from http://www.dsm5.org/Pages/Default.aspx

Arking, D. E., Cutler, D. J., Brune, C. W., Teslovich, T. M., West, K., Ikeda, M., Chakravarti, A. (2008). A common genetic variant in the neurexin superfamily member CNTNAP2 increases familial risk of autism. *Am J Hum Genet, 82*(1), 160-164. doi: 10.1016/j.ajhg.2007.09.015

Autism and Developmental Disabilities Monitoring Network Surveillance Year 2006 (2009). Prevalence of autism spectrum disorders-Autism and Developmental Disabilities

Monitoring Network *MMWR Surveillence Summary*, 58, 1-20. United States, 2006: Centers for Disease Control and Prevention.

Bailey, A., Le Couteur, A., Gottesman, I., Bolton, P., Simonoff, E., Yuzda, E., & Rutter, M. (1995). Autism as a strongly genetic disorder: evidence from a British twin study. *Psychol Med, 25*(1), 63-77.

Baird, G., Simonoff, E., Pickles, A., Chandler, S., Loucas, T., Meldrum, D., & Charman, T. (2006). Prevalence of disorders of the autism spectrum in a population cohort of children in South Thames: the Special Needs and Autism Project (SNAP). *Lancet, 368*(9531), 210-215. doi: 10.1016/s0140-6736(06)69041-7

Benayed, R., Choi, J., Matteson, P. G., Gharani, N., Kamdar, S., Brzustowicz, L. M., & Millonig, J. H. (2009). Autism-associated haplotype affects the regulation of the homeobox gene, ENGRAILED 2. *Biol Psychiatry, 66*(10), 911-917. doi: 10.1016/j.biopsych.2009.05.027

Benayed, R., Gharani, N., Rossman, I., Mancuso, V., Lazar, G., Kamdar, S., . Millonig, J. H. (2005). Support for the homeobox transcription factor gene ENGRAILED 2 as an autism spectrum disorder susceptibility locus. *Am J Hum Genet, 77*(5), 851-868. doi: 10.1086/497705

Bolte, S., Dziobek, I., & Poustka, F. (2009). Brief report: The level and nature of autistic intelligence revisited. *J Autism Dev Disord, 39*(4), 678-682. doi: 10.1007/s10803-008-0667-2

Brown, M. S., & Goldstein, J. L. (1974). Expression of the familial hypercholesterolemia gene in heterozygotes: mechanism for a dominant disorder in man. *Science, 185*(4145), 61-63.

Bucan, M., Abrahams, B. S., Wang, K., Glessner, J. T., Herman, E. I., Sonnenblick, L. I., Hakonarson, H. (2009). Genome-wide analyses of exonic copy number variants in a family-based study point to novel autism susceptibility genes. *PLoS Genet, 5*(6), e1000536. doi: 10.1371/journal.pgen.1000536

Campbell, D. B., Li, C., Sutcliffe, J. S., Persico, A. M., & Levitt, P. (2008). Genetic evidence implicating multiple genes in the MET receptor tyrosine kinase pathway in autism spectrum disorder. *Autism Res, 1*(3), 159-168. doi: 10.1002/aur.27

Campbell, D. B., Sutcliffe, J. S., Ebert, P. J., Militerni, R., Bravaccio, C., Trillo, S., . . . Levitt, P. (2006). A genetic variant that disrupts MET transcription is associated with autism. *Proc Natl Acad Sci U S A, 103*(45), 16834-16839. doi: 10.1073/pnas.0605296103

Carney, R. M., Wolpert, C. M., Ravan, S. A., Shahbazian, M., Ashley-Koch, A., Cuccaro, M. L., . . . Pericak-Vance, M. A. (2003). Identification of MeCP2 mutations in a series of females with autistic disorder. *Pediatr Neurol, 28*(3), 205-211.

Chen, X., Li, X., Wang, P., Liu, Y., Zhang, Z., Zhao, G., Kong, X. (2010). Novel association strategy with copy number variation for identifying new risk Loci of human diseases. *PLoS One, 5*(8), e12185. doi: 10.1371/journal.pone.0012185

Cheng, Y., Sudarov, A., Szulc, K. U., Sgaier, S. K., Stephen, D., Turnbull, D. H., & Joyner, A. L. (2010). The Engrailed homeobox genes determine the different foliation patterns in the vermis and hemispheres of the mammalian cerebellum. *Development, 137*(3), 519-529. doi: 10.1242/dev.027045

Cisse, M., Halabisky, B., Harris, J., Devidze, N., Dubal, D. B., Sun, B., Mucke, L. (2011). Reversing EphB2 depletion rescues cognitive functions in Alzheimer model. *Nature, 469*(7328), 47-52. doi: 10.1038/nature09635

Conciatori, M., Stodgell, C. J., Hyman, S. L., O'Bara, M., Militerni, R., Bravaccio, C., .Persico, A. M. (2004). Association between the HOXA1 A218G polymorphism and increased head circumference in patients with autism. *Biol Psychiatry, 55*(4), 413-419. doi: 10.1016/j.biopsych.2003.10.005

Crespi, B., & Badcock, C. (2008). Psychosis and autism as diametrical disorders of the social brain. *Behav Brain Sci, 31*(3), 241-261; discussion 261-320. doi: 10.1017/s0140525x08004214

Curtis, A. R., Headland, S., Lindsay, S., Thomas, N. S., Boye, E., Kamakari, S., *et al.* (1993). X chromosome linkage studies in familial Rett syndrome. *Hum Genet, 90*(5), 551-555.

Danielsson, S., Gillberg, I. C., Billstedt, E., Gillberg, C., & Olsson, I. (2005). Epilepsy in young adults with autism: a prospective population-based follow-up study of 120 individuals diagnosed in childhood. *Epilepsia, 46*(6), 918-923. doi: 10.1111/j.1528-1167.2005.57504.x

Dickson, S. P., Wang, K., Krantz, I., Hakonarson, H., & Goldstein, D. B. (2010). Rare variants create synthetic genome-wide associations. *PLoS Biol, 8*(1), e1000294. doi: 10.1371/journal.pbio.1000294

Eagleson, K. L., Campbell, D. B., Thompson, B. L., Bergman, M. Y., & Levitt, P. (2011). The autism risk genes MET and PLAUR differentially impact cortical development. *Autism Res, 4*(1), 68-83. doi: 10.1002/aur.172

El-Fishawy, P., & State, M. W. (2010). The genetics of autism: key issues, recent findings, and clinical implications. *Psychiatr Clin North Am, 33*(1), 83-105. doi: 10.1016/j.psc.2009.12.002

Ferreira, M. A., O'Donovan, M. C., Meng, Y. A., Jones, I. R., Ruderfer, D. M., Jones, L., Craddock, N. (2008). Collaborative genome-wide association analysis supports a role for ANK3 and CACNA1C in bipolar disorder. *Nat Genet, 40*(9), 1056-1058. doi: 10.1038/ng.209

Fombonne, E. (2009). Epidemiology of pervasive developmental disorders. *Pediatr Res, 65*(6), 591-598. doi: 10.1203/PDR.0b013e31819e7203

Franz, D. N. (1998). Diagnosis and management of tuberous sclerosis complex. *Semin Pediatr Neurol, 5*(4), 253-268.

Gai, X., Xie, H. M., Perin, J. C., Takahashi, N., Murphy, K., Wenocur, A. S., .White, P. S. (2011). Rare structural variation of synapse and neurotransmission genes in autism. *Mol Psychiatry*. doi: 10.1038/mp.2011.10

Gepner, B., & Feron, F. (2009). Autism: a world changing too fast for a mis-wired brain? *Neurosci Biobehav Rev, 33*(8), 1227-1242. doi: 10.1016/j.neubiorev.2009.06.006

Glessner, J. T., & Hakonarson, H. (2009). Common variants in polygenic schizophrenia. *Genome Biol, 10*(9), 236. doi: 10.1186/gb-2009-10-9-236

Glessner, J. T., Reilly, M. P., Kim, C. E., Takahashi, N., Albano, A., Hou, C., Hakonarson, H. (2010). Strong synaptic transmission impact by copy number variations in schizophrenia. *Proc Natl Acad Sci U S A, 107*(23), 10584-10589. doi: 10.1073/pnas.1000274107

Glessner, J. T., Wang, K., Cai, G., Korvatska, O., Kim, C. E., Wood, S., Hakonarson, H. (2009). Autism genome-wide copy number variation reveals ubiquitin and neuronal genes. *Nature, 459*(7246), 569-573. doi: 10.1038/nature07953

Guilmatre, A., Dubourg, C., Mosca, A. L., Legallic, S., Goldenberg, A., Drouin-Garraud, V., Campion, D. (2009). Recurrent rearrangements in synaptic and neurodevelopmental genes and shared biologic pathways in schizophrenia, autism, and mental retardation. *Arch Gen Psychiatry, 66*(9), 947-956. doi: 10.1001/archgenpsychiatry.2009.80

Hagerman, R. J., Berry-Kravis, E., Kaufmann, W. E., Ono, M. Y., Tartaglia, N., Lachiewicz, A., Tranfaglia, M. (2009). Advances in the treatment of fragile X syndrome. *Pediatrics, 123*(1), 378-390. doi: 10.1542/peds.2008-0317

Happe, F., Ronald, A., & Plomin, R. (2006). Time to give up on a single explanation for autism. *Nat Neurosci, 9*(10), 1218-1220. doi: 10.1038/nn1770

Harris, S. W., Hessl, D., Goodlin-Jones, B., Ferranti, J., Bacalman, S., Barbato, I., Hagerman, R. J. (2008). Autism profiles of males with fragile X syndrome. *Am J Ment Retard, 113*(6), 427-438.

International Molecular Genetic Study of Autism Consortium (IMGSAC). A genomewide screen for autism: strong evidence for linkage to chromosomes 2q, 7q, and 16p. (2001). *Am J Hum Genet, 69*(3), 570-581. doi: 10.1086/323264

Kim, H. G., Kishikawa, S., Higgins, A. W., Seong, I. S., Donovan, D. J., Shen, Y., Gusella, J. F. (2008). Disruption of neurexin 1 associated with autism spectrum disorder. *Am J Hum Genet, 82*(1), 199-207. doi: 10.1016/j.ajhg.2007.09.011

Lai, C. S., Fisher, S. E., Hurst, J. A., Vargha-Khadem, F., & Monaco, A. P. (2001). A forkhead-domain gene is mutated in a severe speech and language disorder. *Nature, 413*(6855), 519-523. doi: 10.1038/35097076

Lauritsen, M. B., Pedersen, C. B., & Mortensen, P. B. (2005). Effects of familial risk factors and place of birth on the risk of autism: a nationwide register-based study. *J Child Psychol Psychiatry, 46*(9), 963-971. doi: 10.1111/j.1469-7610.2004.00391.x

Levinson, D. F., Duan, J., Oh, S., Wang, K., Sanders, A. R., Shi, J., Gejman, P. V. (2011). Copy Number Variants in Schizophrenia: Confirmation of Five Previous Findings and New Evidence for 3q29 Microdeletions and VIPR2 Duplications. *Am J Psychiatry, 168*(3), 302-316. doi: 10.1176/appi.ajp.2010.10060876

Levy, S. E., Mandell, D. S., & Schultz, R. T. (2009). Autism. *Lancet, 374*(9701), 1627-1638. doi: 10.1016/s0140-6736(09)61376-3

Li, W., Cui, Y., Kushner, S. A., Brown, R. A., Jentsch, J. D., Frankland, P. W., Silva, A. J. (2005). The HMG-CoA reductase inhibitor lovastatin reverses the learning and attention deficits in a mouse model of neurofibromatosis type 1. *Curr Biol, 15*(21), 1961-1967. doi: 10.1016/j.cub.2005.09.043

Lifton, R. P., Gharavi, A. G., & Geller, D. S. (2001). Molecular mechanisms of human hypertension. *Cell, 104*(4), 545-556.

Ma, D., Salyakina, D., Jaworski, J.M., Konidari, I., Whitehead, P.L., Andersen, A.N., …Pericak-Vance MA (2009). Ann. Hum. Genet. *73*, 263–273.

Mandy, W. P., & Skuse, D. H. (2008). Research review: What is the association between the social-communication element of autism and repetitive interests, behaviours and activities? *J Child Psychol Psychiatry, 49*(8), 795-808. doi: 10.1111/j.1469-7610.2008.01911.x

Marchetto, M. C., Carromeu, C., Acab, A., Yu, D., Yeo, G. W., Mu, Y., Muotri, A. R. (2010). A model for neural development and treatment of Rett syndrome using human induced pluripotent stem cells. *Cell, 143*(4), 527-539. doi: 10.1016/j.cell.2010.10.016

McClellan, J., & King, M. C. (2010). Genetic heterogeneity in human disease. *Cell, 141*(2), 210-217. doi: 10.1016/j.cell.2010.03.032

Morrow, E. M., Yoo, S. Y., Flavell, S. W., Kim, T. K., Lin, Y., Hill, R. S., Walsh, C. A. (2008). Identifying autism loci and genes by tracing recent shared ancestry. *Science, 321*(5886), 218-223. doi: 10.1126/science.1157657

Mottron, L., Dawson, M., Soulieres, I., Hubert, B., & Burack, J. (2006). Enhanced perceptual functioning in autism: an update, and eight principles of autistic perception. *J Autism Dev Disord, 36*(1), 27-43. doi: 10.1007/s10803-005-0040-7

Muhle, R., Trentacoste, S. V., & Rapin, I. (2004). The genetics of autism. *Pediatrics, 113*(5), e472-486.

Muller, R. A., Shih, P., Keehn, B., Deyoe, J. R., Leyden, K. M., & Shukla, D. K. (2011). Underconnected, but How? A Survey of Functional Connectivity MRI Studies in Autism Spectrum Disorders. *Cereb Cortex*. doi: 10.1093/cercor/bhq296

Muotri, A. R., Marchetto, M. C., Coufal, N. G., Oefner, R., Yeo, G., Nakashima, K., & Gage, F. H. (2010). L1 retrotransposition in neurons is modulated by MeCP2. *Nature, 468*(7322), 443-446. doi: 10.1038/nature09544

Newschaffer, C. J., Croen, L. A., Daniels, J., Giarelli, E., Grether, J. K., Levy, S. E., . Windham, G. C. (2007). The epidemiology of autism spectrum disorders. *Annu Rev Public Health, 28*, 235-258. doi: 10.1146/annurev.publhealth.28.021406.144007

O'Roak, B. J., & State, M. W. (2008). Autism genetics: strategies, challenges, and opportunities. *Autism Res, 1*(1), 4-17. doi: 10.1002/aur.3

Osborne, J. P., Fryer, A., & Webb, D. (1991). Epidemiology of tuberous sclerosis. *Ann N Y Acad Sci, 615*, 125-127.

Pinto, D., Pagnamenta, A. T., Klei, L., Anney, R., Merico, D., Regan, R., Betancur, C. (2010). Functional impact of global rare copy number variation in autism spectrum disorders. *Nature, 466*(7304), 368-372. doi: 10.1038/nature09146

Piven, J., Palmer, P., Jacobi, D., Childress, D., & Arndt, S. (1997). Broader autism phenotype: evidence from a family history study of multiple-incidence autism families. *Am J Psychiatry, 154*(2), 185-190.

Ramocki, M. B., & Zoghbi, H. Y. (2008). Failure of neuronal homeostasis results in common neuropsychiatric phenotypes. *Nature, 455*(7215), 912-918. doi: 10.1038/nature07457

Redies, C. (2000). Cadherins in the central nervous system. *Prog Neurobiol, 61*(6), 611-648.

Richards, R. I., Holman, K., Kozman, H., Kremer, E., Lynch, M., Pritchard, M., Sutherland, G. R. (1991). Fragile X syndrome: genetic localisation by linkage mapping of two microsatellite repeats FRAXAC1 and FRAXAC2 which immediately flank the fragile site. *J Med Genet, 28*(12), 818-823.

Risch, N., & Merikangas, K. (1996). The future of genetic studies of complex human diseases. *Science, 273*(5281), 1516-1517.

Roberts, T. P., Khan, S. Y., Rey, M., Monroe, J. F., Cannon, K., Blaskey, L., .Edgar, J. C. (2010). MEG detection of delayed auditory evoked responses in autism spectrum disorders: towards an imaging biomarker for autism. *Autism Res, 3*(1), 8-18. doi: 10.1002/aur.111

Rogers, S. J., Wehner, D. E., & Hagerman, R. (2001). The behavioral phenotype in fragile X: symptoms of autism in very young children with fragile X syndrome, idiopathic autism, and other developmental disorders. *J Dev Behav Pediatr, 22*(6), 409-417.

Samaco, R. C., Hogart, A., & LaSalle, J. M. (2005). Epigenetic overlap in autism-spectrum neurodevelopmental disorders: MECP2 deficiency causes reduced expression of UBE3A and GABRB3. *Hum Mol Genet, 14*(4), 483-492. doi: 10.1093/hmg/ddi045

Sebat, J., Lakshmi, B., Malhotra, D., Troge, J., Lese-Martin, C., Walsh, T., Wigler, M. (2007). Strong association of de novo copy number mutations with autism. *Science, 316*(5823), 445-449. doi: 10.1126/science.1138659

Skuse, D. H. (2007). Rethinking the nature of genetic vulnerability to autistic spectrum disorders. *Trends Genet, 23*(8), 387-395. doi: 10.1016/j.tig.2007.06.003

Sousa, I., Clark, T. G., Toma, C., Kobayashi, K., Choma, M., Holt, R., .Monaco, A. P. (2009). MET and autism susceptibility: family and case-control studies. *Eur J Hum Genet, 17*(6), 749-758. doi: 10.1038/ejhg.2008.215

St Pourcain, B., Wang, K., Glessner, J.T., Golding, J., Steer, C., Ring, S.M., Skuse, D.H., Grant, S.F., Hakonarson, H., and Smith, G.D. (2010). Am. J. Psychiatry. Published online July 15, 2010. 10.1176/appi. ajp.2010.09121789.

Vernes, S. C., Newbury, D. F., Abrahams, B. S., Winchester, L., Nicod, J., Groszer, M., Fisher, S. E. (2008). A functional genetic link between distinct developmental language disorders. *N Engl J Med, 359*(22), 2337-2345. doi: 10.1056/NEJMoa0802828

Volkmar, F. R. (2005). *Handbook of Autism and Pervasive Developmental Disorders* (3 ed.). New Jersey: John Wiley & Sons.

Wang, K., Zhang, H., Ma, D., Bucan, M., Glessner, J. T., Abrahams, B. S., Hakonarson, H. (2009). Common genetic variants on 5p14.1 associate with autism spectrum disorders. *Nature, 459*(7246), 528-533. doi: 10.1038/nature07999

Wang K, Dickson SP, Stolle CA, Krantz ID, Goldstein DB, Hakonarson H. Interpretation of association signals and identification of causal variants from genome-wide association studies. Am J Hum Genet. 2010 May 14;86(5):730-42. Epub 2010 Apr 29.

Wang, L., Jia, M., Yue, W., Tang, F., Qu, M., Ruan, Y., .Zhang, D. (2008). Association of the ENGRAILED 2 (EN2) gene with autism in Chinese Han population. *Am J Med Genet B Neuropsychiatr Genet, 147B*(4), 434-438. doi: 10.1002/ajmg.b.30623

Yi, J. J., & Ehlers, M. D. (2005). Ubiquitin and protein turnover in synapse function. *Neuron, 47*(5), 629-632. doi: 10.1016/j.neuron.2005.07.008

Zafeiriou, D. I., Ververi, A., & Vargiami, E. (2007). Childhood autism and associated comorbidities. *Brain Dev, 29*(5), 257-272. doi: 10.1016/j.braindev.2006.09.003

Zhong, H., Serajee, F. J., Nabi, R., & Huq, A. H. (2003). No association between the EN2 gene and autistic disorder. *J Med Genet, 40*(1), e4.

5

Genetic Heterogeneity of Autism Spectrum Disorders

Catherine Croft Swanwick, Eric C. Larsen
and Sharmila Banerjee-Basu
MindSpec, Inc.
United States of America

1. Introduction

Although autism is considered to be one of the most highly heritable psychiatric disorders, molecular mechanisms underlying its pathogenesis remain largely unresolved. A strong genetic component underlying autism spectrum disorders (ASD) has been firmly established from various lines of studies ranging from whole genome scans to genetic association studies. Recent genomic advances have led to steep growth in the number of diverse genetic loci linked to ASD, including candidate genes containing rare or common variants, chromosomal aberrations, and submicroscopic copy number variations. Additionally, autism is consistently associated with a number of single gene mutation disorders such as Fragile X Syndrome. Most genetic variations fail to replicate between studies and populations, further complicating our understanding of ASD disease etiology.

Here we review recent expansion of heterogeneity in the genetic landscape for ASD. First we define the types of genetic risk factors implicated in this disorder. We then comparatively analyze the pools of ASD candidate genes identified as of the end of years 2006 and 2010, profiling both their distribution and molecular function. We highlight bioinformatics tools for ASD which can be used to build and evaluate networks of ASD genes as the number of risk factors grows. Finally, we discuss the impact of genetic heterogeneity on theories of ASD pathogenesis.

2. Genes

In the post-genomic era, continuous identification of new ASD risk factors has rapidly expanded the types of candidate genes implicated in the pathogenesis of this disorder. Until 2003, single gene mutations in ASD were derived from well-characterized genetic syndromes such as Fragile X Syndrome and Rett Syndrome, in which subpopulations of individuals develop autistic symptoms. Later that year, Thomas Bourgeron's group first identified single gene mutations/disruptions in neuroligins in siblings with ASD (Jamain et al., 2003). This seminal work opened up the field of ASD research in two major areas: first, a strong genetic foundation to non-syndromic forms of ASD and, second, a focus on the synaptic model for the disorder. Since then, high throughput genetic studies have rapidly identified additional genetic risk factors, vastly expanding the pool of ASD-linked genes.

Candidate genes for ASD can currently be defined into four distinct sets:

1. *Rare*: genes implicated in rare monogenic forms of ASD. The types of allelic variants within this class include rare polymorphisms and single gene disruptions/mutations directly linked to ASD. Examples include *NRXN1* and *SHANK3*.
2. *Syndromic*: genes implicated in syndromes in which a significant subpopulation develops autistic symptoms. Examples include *FMR1* (Fragile X Syndrome) and *MECP2* (Rett Syndrome).
3. *Association*: genes with common polymorphisms that confer small risk for ASD and have been identified from genetic association studies of ASD derived from unknown cause (known as "idiopathic ASD"). Examples include *MET* and *GABRB1*.
4. *Functional*: genes with functions relevant for ASD biology and not included in any of the other genetic categories. Examples include *CADSP2*, for which knockout mouse models exhibit autistic characteristics, but the gene itself has not been directly tied to known cases of ASD.

Of these four gene categories, *Rare* and *Syndromic* contain the strongest evidence of links to ASD (for review, El-Fishawy & State, 2010). *Association* genes lack replication of their relationship to ASD, and *Functional* genes have no documented direct link to ASD. Over 200 ASD candidate genes have been reported thus far in the scientific literature (Table 1). These genes are distributed at discrete regions throughout the entire genome (Figure 1).

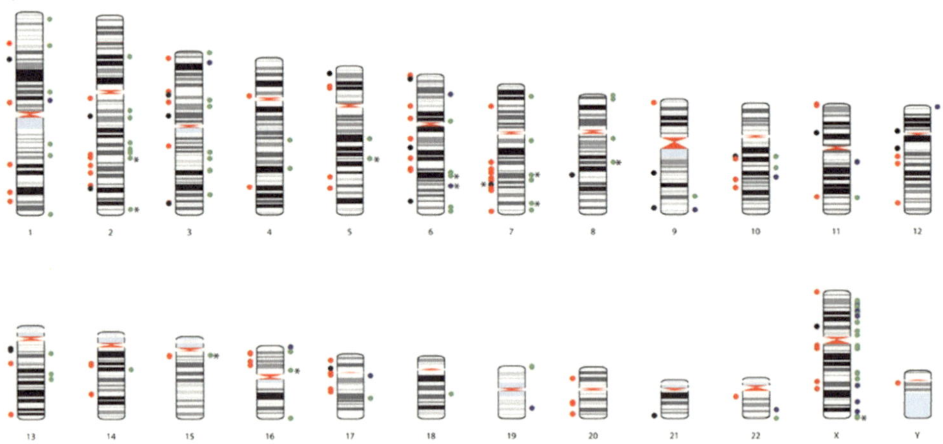

Fig. 1. Ideogram of all currently known ASD candidate genes. To the left of each chromosome are genes that fall within either the *association* category (red) or the *functional* category (black), whereas to the right of each chromosome are genes that fall within either the *rare* mutation category (green) or the *syndromic* category (blue). * = gene also identified by genome-wide association studies.

3. Chromosomes

Microscopically visible large-scale chromosomal rearrangements have long been implicated in the onset and progression of a host of developmental disorders. Deletions of the 15q11-q13 region on the maternal chromosome lead to Angelman syndrome (Williams et al., 2008), whereas the corresponding deletion on the paternal chromosome gives rise to Prader-Willi syndrome (Cassidy & Schwartz, 2009). Deletions, duplications, translocations, and inversions larger than 3 Mb responsible for these and other syndromes have traditionally been identified by microscopic techniques such as karyotyping and, more recently, fluorescent *in situ* hybridization (FISH). In recent years, technological and computational advances have provided researchers with the sensitivity and accuracy to identify structural variation in chromosomes less than 3 Mb in size, which could not have previously been identified by traditional cytogenetic methods such as karyotyping.

Genetic Category	Number of Genes	Genes
Rare	81	ANKRD11, A2BP1*, APC*, ASTN2, AUTS2, BZRAP1, C3orf58, CA6, CACNA1H, CADM1, CENTG2*, CNTN4, CNTNAP2*, CNTNAP5, CXCR3, DIAPH3, DLGAP2, DPP10, DPP6, DPYD, EIF4E, FABP5*, FABP7*, FBXO40, FHIT, FRMPD4, GALNT13, GLRA2, GRPR, HNRNPH2, IL1RAPL1, IMMP2L*, JMJD1C, KCNMA1, KIAA1586, MBD1, MBD3, MBD4, MCPH1, MDGA2, MEF2C, NBEA, NLGN1, NLGN3, NLGN4X, NOS1AP, NRXN1, ODF3L2, OPHN1, OR1C1, PARK2, PCDH9, PCDH10, PCDH19, PDZD4, PLN, PPP1R3F, PSMD10, PTCHD1, RAB39B*, RAPGEF4, RB1CC1, REEP3, RFWD2, RIMS3, RPL10, RPS6KA2, SCN1A, SCN2A, SEZ6L2, SH3KBP1, SHANK2, SHANK3, SLC4A10, SLC9A9, ST7, SUCLG2, TMEM195, TSPAN7, UBE3A*, WNK3
Syndromic	21	ADSL, AGTR2, AHI1*, ALDH5A1, ARX, CACNA1C, CACNA1F, CDKL5, DHCR7, DMD, DMPK, FMR1, MECP2, NF1, NTNG1, PTEN, SLC6A8, SLC9A6, TSC1, TSC2, XPC
Association	84	ABAT, ADA, ADORA2A, ADRB2, AR, ARNT2, ASMT, ATP10A, AVPR1A, C4B, CACNA1G, CCDC64, CDH10, CDH22, CDH9, CTNNA3, CYP11B1, DISC1, DLX1, DLX2, DRD3, EN2, ESR1, ESRRB, FBXO33, FEZF2, FOXP2, FRK, GABRA4, GABRB1, GABRB3, GLO1, GPX1, GRIK2, GRIN2A, GRM8, GSTM1,HLA-A, HLA-DRB1, HOXA1, HRAS, HS3ST5, HSD11B1, HTR1B, HTR3A, HTR3C, INPP1, ITGA4, ITGB3, LAMB1, LRFN5, LRRC1, LZTS2, MACROD2, MARK1, MET, MTF1, MYO16, NOS2A, NPAS2, NRCAM, NRP2, NTRK1, NTRK3, OXTR, PER1, PIK3CG, PITX1, PON1, PRKCB1, PTGS2, RELN, RHOXF1, SLC1A1, SLC25A12, SLC6A4, STK39, SYT17, TDO2, TPH2, UBE2H, VASH1, WNT2
Functional	23	ALOX5AP, ASS, CACNA1D, CADPS2, CBS, CD44, CNR1, DAB1, DAPK1, DCUN1D1, DDX11, EGR2, F13A1, FLT1, ITGB7, MAOA, MAP2, OPRM1, RAI1, ROBO1, SDC2, SEMA5A, TSN

Table 1. Genetic classification of all currently identified ASD candidate genes. (* = gene replicated by independent association studies.)

Of particular interest in the field of submicroscopic structural variants are deletions and duplications collectively categorized as copy number variants. A copy number variant (CNV) is typically defined as a ≥1 kb DNA segment that is present at a differing copy number compared to a reference genome (Feuk et al., 2006). CNVs can either arise *de novo* or be inherited on the maternal and/or paternal chromosome. Much like many single nucleotide polymorphisms, apparently benign CNVs exist in the general population at relatively high frequencies; as such, CNVs that exist in the general population at a rate of 1% or higher are generally described as CNV polymorphisms (Feuk et al, 2006). Submicroscopic copy number variants have come under increased scrutiny in recent years as a potential causative agent in the onset and progression of developmental disorders, including neuropsychiatric disorders such as ASD.

3.1 Copy number variation in autism spectrum disorders

As more syndromes were subsequently shown to be associated with both microscopic and submicroscopic chromosomal structural variation, it became apparent that a subset of patients diagnosed with some of these syndromes also developed ASD or displayed autistic traits. For example, DiGeorge Syndrome (also called Velocardiofacial Syndrome), which is frequently characterized by congenital heart anomalies, palatal abnormalities, immune system deficits and some degree of facial dysmorphism, has been found to result from a ~3 Mb deletion in chromosome 22 (McDonald-McGinn et al., 2005). Individuals diagnosed with this syndrome, also referred to as 22q11.2 deletion syndrome, frequently experience learning disabilities; however, approximately 20% of patients with this syndrome also develop ASD. Given that a subset of patients with syndromes caused by chromosomal structural abnormalities also display autistic traits, as well as the high prevalence of ASD in individuals with cytogenetically visible duplications of the Angelman/Prader-Willi syndrome region (15q11-q13) on the maternal chromosome (Cook Jr. et al., 1997; Schroer et al., 1998), a number of studies in the past decade have focused on identifying submicroscopic structural variants, in particular CNVs, in individuals with ASD and subsequently determining the importance of these variants in disease pathogenesis. In order to more fully ascertain the pathogenic risk associated with copy number variants, only patients with idiopathic cases of ASD have typically been used; patients with mutations in genes previously implicated in ASD, such as the *FMR1* gene, or with gross chromosomal abnormalities have frequently been excluded from these studies.

The advent of genome-wide scanning technologies has enabled researchers to identify and subsequently confirm >1200 potentially pathologically relevant CNVs located within over 490 distinct loci in autistic populations since 2007 (Sebat et al., 2007; Szatmari et al., 2007; Marshall et al., 2008; Cuscó et al., 2009; Glessner et al., 2009; Gregory et al., 2009; van der Zwaag et al., 2009; Pinto et al., 2010; Bremer et al., 2011). Confirmation or validation of a CNV by an independent approach following its discovery is essential not only to remove false positives, but also to more accurately identify the boundaries of a CNV. Validated CNVs in autistic individuals have been located in loci on all 22 somatic chromosomes and the X chromosome (Figure 2).

While many of the CNVs identified by these methods are singletons and require additional replication to more accurately assess their potential role in disease, there are rare, recurring CNVs at particular loci that have been identified across multiple autistic populations that have emerged as strong risk-conferring candidates in ASD pathogenesis. Ten loci that have been identified multiple times in autistic case populations are described in Table 2. Perhaps the most intensely studied of these recurring CNVs, aside from duplications in the 15q11-13 loci, are ~500 kb deletions and duplications that occur at the 16p11.2 locus. A recently published meta-analysis of the 16p11.2 locus in autistic populations discovered that CNVs at the 16p11.2 locus have a prevalence of 0.76%, with deletions occurring approximately twice as frequently with duplications (Walsh & Bracken, 2011). CNVs in autistic individuals have been identified in regions previously associated with other deletion-duplication syndromes, such as the 1q21.1, 22q11.21 and 22q13.33 loci (McDonald-McGinn et al., 2005; Phelan, 2007; Haldeman-Englert & Jewett, 2011). Other strong candidate CNV loci to emerge from genome-wide scanning assays include 2p16.3, 3p26.3, 6q26, 7q11.22, and 15q13.3. In some cases, CNVs at these "hot-spot" loci appear to target genes that have previously been implicated in ASD pathogenesis, such as *NRXN1* (2p16.3), *PARK2* (6q26), and *AUTS2* (7q11.22).

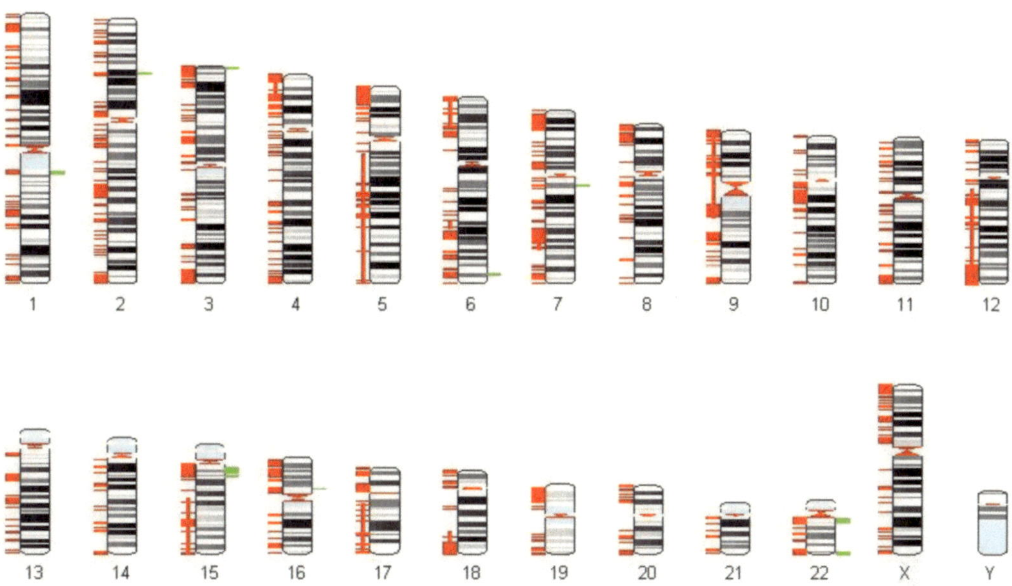

Fig. 2. Validated copy number variants (CNVs) identified in genome-wide scanning arrays from nine published reports (Sebat et al., 2007; Szatmari et al., 2007; Marhall et al., 2008; Cuscó et al., 2009; Glessner et al., 2009; Gregory et al., 2009; van der Zwaag et al., 2009; Pinto et al., 2010; Bremer et al., 2011). The red lines to the left of each chromosome represent the >1200 validated CNVs identified in these studies. Especially long CNVs that overlap smaller CNVs are represented with thinner red lines. The green lines to the right of chromosomes 1, 2, 3, 6, 7, 15, 15, and 22 represent ten rare, recurring CNV loci identified in at least three of the nine aforementioned publications (described in more detail in Table 2).

Increasingly, targeted assays using methods such as quantitative PCR are being used to characterize CNVs at particular loci that have been previously identified by more global scanning approaches, given the relatively high frequency of these CNVs in autistic case cohorts. CNVs are now considered one of the most common, genetic causes of ASD, with 10-20% of ASD cases believed to be the result of submicroscopic deletions and duplications (Miles et al., 2010).

CNV locus	Candidate gene(s)	CNV Type	References
1q21.1		Deletion-Duplication	Szatmari et al., 2007; Gregory et al., 2009; Pinto et al., 2009
2p16.3	*NRXN1*	Deletion-Duplication	Szatmari et al., 2007; Glessner et al., 2009; Pinto et al., 2010; Bremer et al., 2011
3p26.3*	*CNTN4*	Deletion-Duplication	Glessner et al., 2009; Pinto et al., 2010; Bremer et al., 2011
6q26	*PARK2*	Deletion-Duplication	Szatmari et al., 2007; Glessner et al., 2009; Pinto et al., 2010; Bremer et al., 2011
7q11.22	*AUTS2*	Duplication	Cuscó et al., 2009; Glessner et al., 2009, Pinto et al., 2010
15q11-q13	*UBE3A, GABRB3, GABRA5, GABRG3*	Duplication	Sebat et al., 2007; Marshall et al., 2008; Glessner et al., 2009; Pinto et al., 2010
15q13.3*	*CHRNA7*	Deletion-Duplication	Gregory et al., 2009; Pinto et al., 2010; Bremer et al., 2011
16p11.2		Deletion-Duplication	Sebat et al., 2007; Marshall et al., 2008; Glessner et al., 2009; Pinto et al., 2010; Bremer et al., 2011
22q11.21		Deletion-Duplication	Szatmari et al., 2007; Marshall et al., 2008; Glessner et al., 2009; Pinto et al., 2010; Bremer et al., 2011
22q13.33	*SHANK3*	Deletion-Duplication	Szatmari et al., 2007; Marshall et al., 2008; Glessner et al., 2009; van der Zwaag et al., 2009; Pinto et al., 2010; Bremer et al., 2011

Table 2. Examples of loci in which rare, recurring, potentially risk-conferring CNVs are frequently observed in autistic populations. Potential candidate genes within the locus, the type of CNV that targets each locus, and the articles in which these CNV loci were identified are included. *, CNV locus which overlaps with adjacent loci in at least one of the publications listed.

A more detailed analysis of the nine published research articles used to construct the ideogram in Figure 2 reveals that, while the percentage of previously unidentified CNV loci has steadily declined since 2007, new CNV loci still constitute a very high percentage of the total CNV loci identified and validated in these studies (Table 3). Therefore, while recurring CNVs such as 16p11.2 and others continue to be observed across multiple autistic populations and CNV studies, novel CNVs in autistic populations are still being identified, indicating that there are likely multiple potential targets for the pathogenic properties of CNVs throughout the human genome. It is likely that other novel CNVs in autistic individuals have not yet been identified, and as such their identification will shed new light on the pathways adversely affected in ASD.

Year	2007	2008	2009	2010-11
# of published reports	2	1	4	2
# total CNV loci identified	98	32	56	396
# previously unidentified CNV loci	97	27	47	320
% of CNV loci previously unidentified	**98.98**	**84.38**	**83.93**	**80.81**

Table 3. Analysis of the nine papers used to construct the ideogram in Figure 2. while the overall percentage of previously unidentified CNV loci has decreased from year to year, novel CNV loci still constitute the majority of the total CNV loci that have been identified and confirmed in these studies.

3.2 Risk-conferring vs. benign copy number variants

Although advances in genome-wide and targeted scanning assays have enabled researchers to discover potentially risk-conferring CNVs in autistic individuals, significant issues remain in the determination of which CNVs are pathologically relevant or benign in nature. This is of particular importance in terms of potentially using genetic screening for risk-conferring CNVs as a tool to assess the risk of ASD in unborn children. The diagnostic accuracy of such a screening protocol would be entirely dependent on knowing which CNVs would confer the greatest potential risks for ASD pathogenesis. In order to distinguish between risk-conferring and benign CNVs in an autistic population, a comparison must be made between both the existence and frequency of CNVs between affected and unaffected individuals. To account for possible genetic differences between ethnic groups, it is critical that a control population of comparable size and ethnic background be included in any CNV study. For example, CNVs at loci thought to confer a high risk of ASD susceptibility, such as deletions and duplications at the 16p11.2 locus, have also been identified in healthy individuals, although at a much lower frequency than in autistic populations. Given the increased frequency of CNVs at the 16p11.2 loci in autistic populations versus control populations, CNVs at this region remain classified as high risk-conferring CNVs. In addition, there are online databases such as the Database of Genomic Variants (http://projects.tcag.ca/variation/) and the Copy Number Variant resource at the Children's Hospital of Philadelphia (http://cnv.chop.edu/) available that describe previously identified CNVs in healthy individuals. These tools provide a means to further filter out likely benign CNVs from autistic case studies and enrich for potentially pathogenic variants. However, it should be noted that seemingly benign CNVs may be involved in more subtle phenotypes in autistic individuals when occurring in combination with other factors. Likewise, additional meta-analysis studies of CNV loci across multiple published autistic populations, such as that described for the 16p11.2 locus, will be required to compare frequencies of CNV a in order to more fully determine the global risk potential associated with any given CNV at a particular locus.

3.3 De novo vs. inherited copy number variants

As previously stated, CNVs can either arise *de novo*, or be inherited from the mother and/or father. Considerable interest has been placed in the pathogenic importance of *de novo* CNVs as a cause of ASD compared to inherited variants, especially within the context of sporadic vs. familial ASD cases. Indeed, some studies have found that the rate of *de novo* CNVs is higher in sporadic cases compared to familial cases (Sebat et al., 2007; Marshall et al., 2008), while Bremer et al. (2011) found that the rate of rare inherited CNVs was higher in familial cases compared to sporadic cases. These findings would suggest that *de novo* CNVs are predominantly responsible for ASD in sporadic cases, whereas inherited CNVs are primarily responsible for familial cases of ASD. However, Pinto et al. (2010) found no significant difference between the frequencies of *de novo* CNVs in sporadic vs. familial cases. It has been reported that validated de novo CNVs strongly associate with ASD (Sebat et al., 2007). However, there is no firm evidence that *de novo* CNVs confer a higher probability or severity of disease than inherited variants. On the other hand, the dynamics of CNV inheritance and subsequent susceptibility to ASD has its own issues: an autistic individual with a potential risk-conferring CNV may inherit that CNV from a parent who fails to exhibit autistic traits; an autistic individual may have unaffected siblings who have likewise inherited the identical CNV; or one affected sibling in a multiple family may have a risk-conferring CNV, whereas other affected siblings may not.

3.4 Copy number variation and phenotypic heterogeneity

Detailed studies attempting to correlate genotype with phenotype have demonstrated that there is significant phenotypic heterogeneity between individuals with CNVs at a particular chromosomal locus, both in terms of disease presence and severity of disease. Studies in autistic populations containing CNVs at the 15q13.3 (Miller et al., 2009; Ben-Shachar et al., 2009) and 16p11.2 (Fernandez et al., 2010) loci, for example, have shown that autistic phenotypes, such as the extent of facial dysmorphism and the extent of intellectual disability, can vary from one patient to the next with the same CNV. One model that has been designed to address some of the issues as to how CNVs contribute to ASD states that certain CNVs at particular loci increase the susceptibility of an individual to developing an ASD based on a "threshold" of disease severity (Cook & Scherer, 2008). Chief among these high susceptibility CNVs are maternal duplications at 15q11-q13, deletions at 16p11.2, and deletions at the loci encoding for cell adhesion proteins such as neuroligins. Other rare recurring CNVs that have been identified in autistic populations may confer a lower overall risk of ASD pathogenesis, or a decreased severity of disease, such as CNVs at 1q21.1, 2p16.3, and 22q11.21. However, even these CNVs can result in the onset of ASD, or more severe disease phenotypes, when in combination with other genetic and non-genetic factors. These genetic factors may include additional CNVs (indeed, many autistic individuals have more than one CNV within their genome) or single gene mutations, such as those described elsewhere in this chapter, whereas non-genetic factors can be environmental, sex-related, or epigenetic in nature. Epigenetic regulation of gene expression may be of particular importance with regards to phenotypic heterogeneity in autistic individuals with 15q11-q13 duplications, as this region contains a number of potentially critical imprinted genes. Further studies involving more detailed analysis of genotype-phenotype correlations in autistic individuals with CNVs will be instrumental in determining the role of CNVs in ASD.

3.5 Mechanism of action of copy number variants

The general mechanism by which a CNV might contribute to ASD pathogenesis remains unclear. The simplest mechanism of action involves gene dosage, by which deletion or duplication of a gene or genes within a particular CNV locus, or the deletion or duplication of gene regulatory elements, subsequently results in altered or disrupted levels of gene product. A deletion at a particular locus might also result in the unmasking of a recessive gene on the corresponding chromosomal locus, which would then be able to elicit a deleterious effect. Such a mechanism might be involved in disease pathogenesis in an autistic individual with a 10 Mb maternally inherited deletion in chromosome 13q and a point mutation in the *DIAPH3* gene on the paternal chromosome (Vorstman et al., 2010). As the proband's unaffected sibling also had the *DIAPH3* mutation, but lacked the corresponding deletion, it is tempting to argue that the maternal deletion unmasked a recessive mutation in the paternal *DIAPH3* gene, and that in turn influenced the onset of ASD in the proband. Given that many CNVs are large enough to include up to 50 or more genes, identifying which genes are of functional relevance in ASD pathogenesis within a particular CNV loci remains a challenging task. Much in the same way that genes that confer susceptibility to ASD have been found to fall within intriguing functional categories, bioinformatic analysis of genes that lie within or adjacent to recurring CNV loci may yield similar results and aid in both identifying new candidate genes and in discovering conserved pathways potentially targeted by copy number variation. Indeed, analysis designed to identify potentially relevant functional pathways containing genes located in copy number variants have been performed (Pinto et al., 2010).

4. Comparative analysis of ASD genes

To analyze recent evolution of the ASD molecular landscape, we profiled ASD genes identified as of the end of years 2006 and 2010. To define pools of ASD candidate genes existing at these time points, we used the ASD database AutDB (www.mindspec.org/autdb.html), a publicly available, curated, web-based, searchable genetic database for ASD created by our laboratory (see Section 5). We then examined the genetic and functional expansion of these gene sets.

4.1 Genetic expansion

To quantify the total number of ASD candidate genes identified as of 2006 and 2010, we sorted existing ASD candidate genes according to year of first publication. We discovered that the total number of ASD candidate genes more than doubled in the past four years: whereas 91 genes were linked to ASD as of 2006, this number rapidly grew to 209 genes in 2010.

To compare genetic distribution within these datasets, we defined ASD candidate genes according to the classification system described in Section 2: rare variants (*Rare*), syndromic genes (*Syndromic*), genes identified by association studies (*Association*), and genes whose functions have been implicated in ASD (*Functional*). We found that expansion of the total ASD gene pool was largely due to steep growth of both *Rare* and *Association* gene sets, with a slight increase in the numbers of identified *Syndromic* and *Functional* genes (Figure 3). Notably, the near quadrupling of the number of *rare* mutations supports the *Rare Allele, Common Disease* as a plausible theory of ASD pathogenesis (see Section 6).

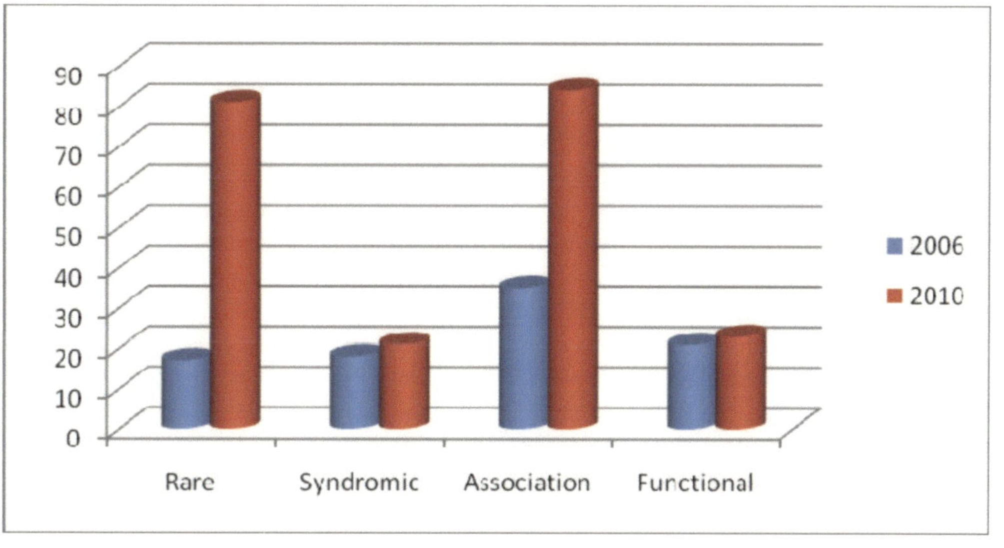

Fig. 3. Genetic distribution of ASD candidate genes identified as of the end of years 2006 and 2010.

4.2 Functional expansion

Recent large-scale ASD studies have used a systems biology approach to translate genetic information into functional maps. For instance, Glessner et al. (2009) showed that ASD-linked genes cluster in synaptic processes such as cell adhesion and ubiquitin-mediated degradation. Additionally, in the largest ASD study performed to date, Pinto et al. (2010) found that genes affected by rare CNVs were enriched in functions such as neuronal development and GTPase/Ras signaling.

To build upon these functional maps, we used a well-known synaptic proteome classification system (Husi et al., 2000), to organize ASD gene sets from 2006 and 2010 into eight broad categories of molecular function, defined by corresponding subcategories:

1. *Cell Adhesion* (cell adhesion molecule, cell adhesion/axon guidance, extracellular matrix, extracellular secreted protein)
2. *Guidance/Outgrowth* (axon guidance, cell migration, cell surface glycoprotein, cytoskeletal remodeling, dendritic spine morphology, animal model evidence)
3. *Neurotransmission* (adaptor protein, G-protein coupled receptors, ligand-gated ion channel, neuromodulator receptor, neuromodulator receptor-associated protein, neuromodulator synthesis, neurotransmitter receptor, neurotransmitter synthesis, presynaptic release, scaffolding protein, sensory receptor, transporter, voltage-gated ion channel, voltage-gated ion channel modulator)
4. *Signaling* (glycosylation, kinase, kinase substrate, phosphatase, proteoglycan, small G-protein or modulator, tyrosine receptor kinase, other signal)
5. *Degradation* (proteasome-related protein, ubiquitin ligase)
6. *Transcription* (circadian protein, cofactor, DNA binding, DNA damage response protein, DNA methylation, estrogen receptor, histone demethylation protein, homeodomain protein, preinitiation complex, purine metabolism, transcription factor)
7. *Translation* (ribosomal protein, RNA binding, RNA metabolism, RNA structure)
8. *Other* (antioxidant, endosome regulation, energy production, fatty acid binding protein, immune system, membrane biosynthesis, mitochondrial carrier protein, mitochondrial targeting protein, oxidation, prostaglandin, unknown function).

The functional distribution of ASD risk genes vastly expanded from 2006 to 2010 (Figure 4). Because *Rare* and *Syndromic* genes contain the strongest links to ASD (see Section 2), we examined this combined "*Rare/Syndromic*" set as one dataset. We comparatively assessed them with *Association* genes as a separate gene set. Both *Rare/Syndromic* and *Association* gene datasets followed the same trend: whereas *Neurotransmission* and *Signaling* were by far the largest functional categories in 2006, the number of genes in all other functional categories increased over the past four years such that all are becoming relatively equalized. The most dramatic increases occurred in *Cell Adhesion*, *Degradation*, *Transcription*, and *Other*.

This functional expansion has led to shifting theories of ASD pathogenesis. In 2006, the largest percentage of ASD susceptibility genes resided in the *Neurotransmission* or *Signaling* categories, supporting specific theories of dysfunction, such as serotonin transport (Cook & Leventhal, 1996). However, rapid expansion of nearly all functional categories throughout 2010 indicates that ASD susceptibility genes are actually widespread in neurobiological function. Such functional expansion supports broad theories of pathogenesis such as the proposed enhancement of brain excitability in ASD (Rubenstein & Merzenich, 2003). Each designated functional category includes neurobiological factors that contribute to brain excitability, reinforcing the idea that mutations in vastly different genes may facilitate similar outcomes in brain function by contributing to shared molecular pathways. Together, accelerated identification of ASD risk genes with widespread neurobiological functions is leading to a convergent model of ASD pathogenesis.

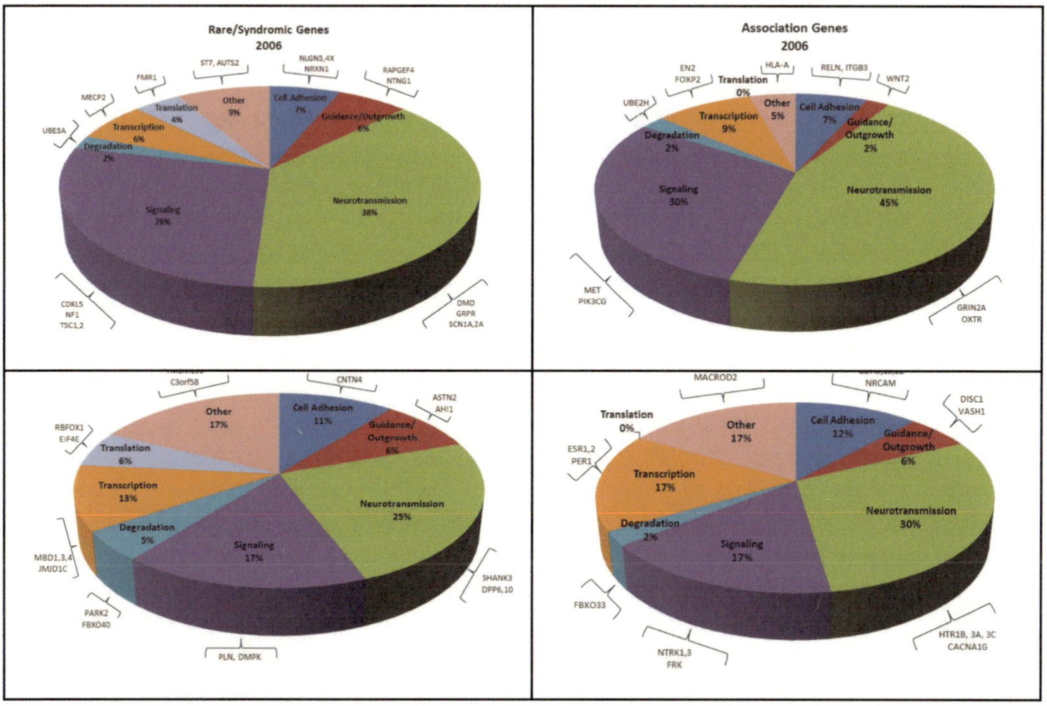

Fig. 4. Functional profile of ASD candidate genes identified as of the end of years 2006 and 2010.

5. Bioinformatics of ASD

The enormous amount of data currently being generated by large-scale genomic studies poses a critical challenge for its storage and analysis. To process this information, bioinformatics tools are becoming increasingly vital to the scientific community. Here we highlight several ASD-related databases which researchers can use to navigate this data and shed insight into the molecular pathways underlying ASD pathogenesis.

5.1 AutDB

Our laboratory created the ASD database AutDB (www.mindspec.org/autdb.html), the first publicly available, curated, web-based, searchable genetic database for ASD (Basu et al., 2009; Kumar et al., 2011). In AutDB, evidence regarding ASD candidate genes is systematically extracted from peer-reviewed, primary scientific literature and manually curated by our researchers. To provide high-resolution view of various components linked to ASD, we developed detailed annotation rules based on the biology of each data type and generated controlled vocabulary for data representation. AutDB is widely used by individual laboratories (Crespi et al., 2010; Elia et al., 2010; Gillis et al., 2010; Toro et al., 2010) and consortiums (Simons Foundation) for understanding genetic bases of ASD.

With a systems biology approach, AutDB integrates various modules encompassing different types of data relevant for ASD:

Human Gene: This original module of AutDB includes all genes whose mutations have been associated or implicated with ASD, together with all risk-conferring candidates associated with these disorders (Basu et al., 2009). ASD-related genes are classified into the four categories described in Section 2: *1) Rare*: genes implicated in rare monogenic forms of ASD; *2) Syndromic*: genes implicated in syndromic forms of ASD where a subpopulation with a specific genetic syndrome develops autistic symptoms; *3) Association*: small risk-conferring candidate genes with common polymorphisms identified from genetic association studies in idiopathic ASD; and, *4) Functional*: candidates genes with functions relevant for ASD biology, not covered by any of the previous genetic categories. All known ASD-specific mutations at the DNA sequence level will be available by late 2011.

Animal Model: This module provides a comprehensive collection of all mouse models linked to ASD (Kumar et al., 2011). The core behavioral features of ASD involve higher order human brain functions like social interactions and communications, which can only be approximated in animal models, so the annotation strategy for this module includes four broad areas: 1) core behavioral features of ASD, 2) ASD-related traits such as seizures and circadian rhythms that are heritable and more easily quantified in animal models; 3) neuroanatomical features, and 4) molecular profiles. To this end, we developed PhenoBase, a classification table for systematically annotating models with controlled vocabulary. PhenoBase contains 16 major categories and >100 standardized phenotype terms.

Protein Interaction (PI): This module serves as a repository for all known protein-protein interactions of ASD candidate genes. It documents five major types of direct interactions: 1) protein binding, 2) promoter binding, 3), RNA binding, 4) protein modification, and 5) direct regulation. One of the newest additions to AutDB, a beta version of this module was released in April 2011, with a full version scheduled for release in late 2011. Its content is envisioned to have immediate application for network biology analysis of molecular pathways involved in ASD pathogenesis.

Copy Number Variant (CNV): This module is a comprehensive, up-to-date reference for all known copy number variants (CNVs) implicated in ASD (see Section 3). It originates from a multi-level annotation model including data such as chromosomal location, size, and relevance to ASD. Like the PPI module, a beta version of the CNV module was released in May 2011, with a full version scheduled for release in late 2011.

5.2 ASD Chromosome Rearrangement Database

The ASD Chromosome Rearrangement Database (http://projects.tcag.ca/ASD/) is a web-based, searchable genetic database hosted by The Centre for Applied Genomics at the Hospital for Sick Children in Toronto, Canada (Marshall et al., 2008). The ASD Chromosome Rearrangement Database provides information not only on submicroscopic CNVs, that have been identified by microarray studies, but also data on microscopic structural variants identified by cytogenetic studies. This database is updated both from published research articles and in-house experimental results.

5.3 ASD Genetic Database

The ASD Genetic Database (http://wren.bcf.ku.edu/) is another web-based, searchable genetic database developed by researchers at the University of Kansas (Matuszek & Talabizadeh, 2009). Much like AutDB and the ASD Chromosome Rearrangement Database, the ASD Genetic Database provides information on genes and CNVs believed to impart susceptibility to ASD. However, this database also includes information on known non-coding RNAs and chemically-induced fragile sites in the human genome. Non-coding RNAs, such as microRNAs, have come under increased scrutiny with regards to their potential pathogenic role in ASD. For example, the 15q11-q13 region contains a number of small nucleolar RNAs (snoRNAs). Duplication of a region of mouse chromosome 7 that has conserved linkage with human chromosome 15q11-q13 in mouse model of ASD resulted in overexpression of the snoRNA MBII52 (the mouse ortholog of the human snoRNA HBII52), which could potentially alter serotonergic signaling and contribute in part to the autistic traits exhibited by these mice (Nakatani et al., 2009). Spontaneous breakage during DNA replication at rare chromosomal fragile sites may also play a role in the pathogenesis of neuropsychiatric disorders such as ASD. The chromosomal fragile site FRAXA has been implicated in fragile X syndrome, and other fragile sites have been identified that associate with ASD, such as FRA2B, FRA6A, and FRA13A (Smith et al., 2010).

6. Discussion

6.1 Rare *vs.* common alleles

At the beginning of this decade, few single mutations for ASD had been identified. As of 2003, single mutations in only two genes were known: neuroligins 3 and 4, published in a single report (Jamain et al., 2003). This led to predominance of the *Common Allele Common Disease* theory, which proposes that ASD is caused by combined effects of multiple common polymorphisms.

However, evidence from two recent major studies led to the emergence of an alternative *Rare Allele Common Disease* theory for ASD pathogenesis. First, comparative genomic hybridization with subsequent confirmation showed a strong association between *de novo* **CNVs** mutations and ASD (Sebat et al., 2007). Second, homozygosity mapping identified numerous single gene mutations in families with ASD (Morrow et al., 2008).

According to the *Rare Allele Common Disease* theory, the genetics underlying complex neuropsychiatric disorders such as ASD is highly heterogeneous. It proposes that ASD is caused by numerous rare, highly penetrant mutations that may even by caused by "private mutations" specific to individual families; a similar theory has been proposed to explain the genetic complexity of schizophrenia (McClellan et al., 2007). The identification of rare variants has more than quadrupled in the past four years (see Section 4), lending credibility to this theory.

At present, it appears that the *Rare Allele Common Disease* theory is a highly relevant genetic paradigm for ASD and other complex disorders. A few recent papers have identified common variants associated with ASD (Campbell et al., 2006; Wang et al., 2009; Weiss et al., 2009; Anney et al., 2010), but these mutations are still far outnumbered by known rare single gene mutations. With increased availability of various types of sequencing technologies, it is projected that additional rare mutations/variations will be discovered or validated rapidly in upcoming years, making clinical genomics of ASD an option for affected families.

6.2 Prioritization of genetic ASD risk factors

In future, ASD risk genes should be prioritized based on careful definitions at both genetic and functional levels. High priority genes should show evidence for replication or participate in a molecular pathway exhibiting multiple ASD-linked mutations. Examples include the cell adhesion molecule *CNTNAP2*, a neurexin family member in which both common and rare variants have been associated with ASD (Arking et al., 2008). *CNTNAP2* is also regulated by *FOXP2*, a candidate ASD gene highly relevant for human language development (Gong et al., 2004). Additionally, the role of synaptic scaffolding proteins in ASD has been strengthened by recent identification of recurrent mutations in *SHANK2* (Berkel et al., 2010; Pinto et al., 2010). Furthermore, if multiple genes contribute to syndromic ASD, each gene should only be considered high priority when accompanied by a documented direct genetic link to ASD.

CNVs should likewise be prioritized based on a number of factors. A high risk-conferring structural variant should not only display a high prevalence in autistic populations, but also an enrichment in autistic populations compared to control populations. Meta-analysis studies, such as that previously described for the 16p11.2 locus in multiple autistic case studies (Walsh & Bracken, 2011), will aid greatly in determining which CNVs meet these criteria. Furthermore, a high priority CNV locus should either contain a gene or genes, or the regulatory elements for a gene or genes, which demonstrate potential participation in a molecular pathway exhibiting multiple ASD-linked single gene mutations. CNV loci containing genes that have already been associated with increased risk of autism, such as 2p16.3 (*NRXN1*), are of particular interest in this regard.

6.3 Synaptic theory of ASD

A hypothesis for ASD as a synaptic disorder is well recognized, largely based on strong evidence from rare mutations in neuroligins, neurexins and *SHANK3* (Bourgeron, 2009). Rapid expansion of the ASD risk gene pool has supported this synaptic theory of ASD by identifying rare mutations in numerous additional synapse-related genes, including *SHANK2* (Berkel et al., 2010; Pinto et al., 2010) and *PTCHD1* (Marshall et al., 2008; Noor et al., 2010; Pinto et al., 2010). Additionally, functional maps generated from large-scale studies of ASD have enriched this synaptic hypothesis of ASD, identifying categories ranging from cell adhesion and ubiquitin-mediated degradation (Glessner et al., 2009) to neuronal development and GTPase/Ras signaling (Pinto et al., 2010).

Our functional profile of all ASD candidate genes identified as of 2010 supports this synaptic hypothesis (see Section 4). The majority of ASD-linked genes function in synaptic processes such as cell adhesion, guidance/outgrowth, neurotransmission, signaling, degradation, transcription, and translation. A smaller fraction of ASD risk genes possessed unknown functions or "Other" non-synaptic functions. Examples of synaptically enriched ASD gene functions are modeled in Figure 5.

7. Conclusion

In conclusion, the broadened molecular landscape for ASD suggests that an integrated approach is required to understand functional pathways underlying ASD. An unbiased view of ASD risk gene datasets emphasizes the importance of overall synaptic networks for human cognition. Higher order functions require efficient information processing, and mutations in any synaptic component could lead to the range of impairments present in ASD. Future spatiotemporal mapping of ASD gene expression patterns may provide clues to how shared susceptibility genes give rise to different forms of ASD. Moreover, identification of new ASD-associated genes using advanced techniques like deep sequencing will increasingly sharpen our functional understanding of ASD synapse biology.

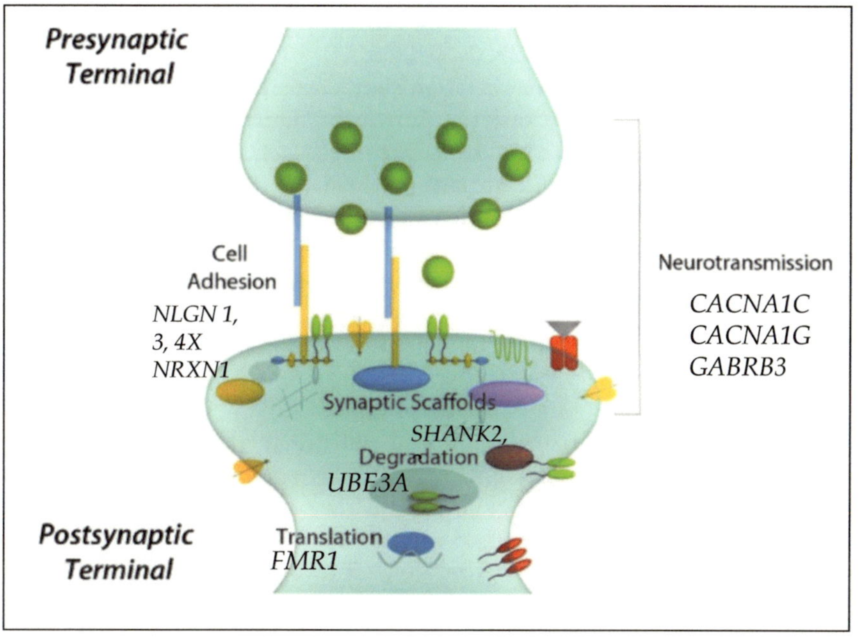

Fig. 5. Synaptic enrichment of ASD candidate genes. ASD risk genes are concentrated in numerous synaptic functions, including cell adhesion (examples: *NLGN1, NLGN3, NLGN4X, NRXN1*), synaptic scaffolds (examples: SHANK2-3), degradation (example: *UBE3A*), translation (example: *FMR1*), and neurotransmission (examples: *CACNA1C, CACNA1G, GABRB3*).

8. Acknowledgements

The authors would like to thank the other members of Mindspec, Inc. (Ajay Kumar, Cynthia Soderblom, Nicole Johnson, Rachna Wadhawan, Rainier Rodriguez, and Sue Spence), as well as the Simons Foundation. AutDB is licensed to the Simons Foundation as SFARI Gene.

9. References

Anney, R., et al. (2010). A genome-wide scan for common alleles affecting risk for autism. *Human Molecular Genetics*, Vol. 19, No. 20, pp. 4072-4082, ISSN 0964-6906.

Arking, D.E., et al., (2008). A common genetic variant in the neurexin superfamily member CNTNAP2 increases familial risk of autism. *American Journal of Human Genetics*, Vol. 82, No. 1 (January 2008), pp. 160-164, ISSN 0002-9297.

Basu, S.N.; Kollu, R. & Banerjee-Basu, S. (2009). AutDB: a gene reference resource for autism research. *Nucleic Acids Research* Vol. 37, pp. D832-D836, ISSN 0305-1048.

Ben-Shachar, S., et al. (2009). Microdeletion 15q13.3: a locus with incomplete penetrance for autism, mental retardation, and psychiatric disorders. *Journal of Medical Genetics*, Vol. 46, pp. 382-388.

Berkel, S., et al. (2010). Mutations in the SHANK2 synaptic scaffolding gene in autism spectrum disorder and mental retardation. *Nature Genetics*, Vol. 42, No. 6 (June 2010), pp. 489-491, ISSN 1061-4036.

Bourgeron, T. (2009). A synaptic trek to autism. *Current Opinion in Neurobiology,* Vol. 19, No. 2 (April 2009), pp. 231-234, ISSN 0959-4388.

Bremer, A., et al. (2011). Copy number variation characteristics in subpopulations of patients with autism spectrum disorders. *American Journal of Human Genetics Part B Neuropsychiatric Genetics,* Vol. 156, No. 2, pp. 115-124, ISSN 1552-4841.

Campbell, D.B., et al. (2006). A genetic variant that disrupts MET transcription is associated with autism. *Proceedings of the National Academy of Sciences U. S. A.,* Vol. 103, No. 45 (November 7, 2006), pp. 16834-16839, ISSN 0027-8424.

Cassidy, S.B. & Schwartz, S. (2009). Prader-Willi syndrome, In: *GeneReviews,* R.A. Pagon et al. (Ed.), University of Washington, Seattle, Retrieved from http://www.ncbi.nlm.nih.gov/books/NBK1116/

Cook, E.H. & Leventhal, B.L. (1996). The serotonin system in autism. *Current Opinion in Pediatrics,* Vol. 8, No. 4 (August 1996), pp. 348-354, ISSN 1040-8703.

Cook Jr., E.H., et al. (1997). Autism or atypical autism in maternally but not paternally derived proximal 15q duplication. *American Journal of Human Genetics,* Vol. 60, No. 4 (April 1997), pp. 928-934, ISSN 0002-9297.

Cook Jr., E.H. & Scherer, S.W. (2008). Copy-number variations associated with neuropsychiatric conditions. *Nature,* Vol. 455, No. 7215 (October 16, 2008), pp. 919-923, ISSN 0028-0836.

Cuscó, I., et al. (2009). Autism-specific copy number variants further implicate the phosphatidylinositol signaling pathway and the glutamatergic synapse in the etiology of the disorder. *Human Molecular Genetics,* Vol. 18, No. 10 (May 15, 2009), pp. 1795-1804, ISSN 0964-6906.

El-Fishawy, P. & State, M.W. (2010). The genetics of autism: key issues, recent findings, and clinical implications. *Psychiatric Clinics of North America,* Vol. 33, No. 1 (March 2010), pp. 83-105, ISSN 0193-953X.

Fernandez, B.A., et al. (2010). Phenotypic spectrum associated with *de novo* and inherited deletions and duplications at 16p11.2 in individuals ascertained for diagnosis of autism spectrum disorder. *Journal of Medical Genetics,* Vol. 47, No. 3 (March 2010), pp. 195-203, ISSN 0022-2593.

Feuk, L.; Carson, A.R. & Scherer, S.W. (2006). Structural variation in the human genome. *Nature Reviews Genetics,* Vol. 7, No. 2 (February 2006), pp. 85-97, ISSN 1471-0056.

Glessner, J.T., et al. (2009). Autism genome-wide copy number variation reveals ubiquitin and neuronal genes. *Nature,* Vol. 459, No. 569 (May 28, 2009), pp. 569-573, ISSN 0028-0836.

Gong, X., et al. (2004). Association between the FOXP2 gene and autistic disorder in Chinese population. *American Journal of Human Genetics Part B Neuropsychiatric Genetics,* Vol. 127B, No. 1 (May 15, 2004), pp. 113-116, ISSN 1552-4841.

Gregory, S.G., et al. (2009). Genomic and epigenetic evidence for oxytocin receptor deficiency in autism. *BMC Medicine,* Vol. 7, No. 62 (October 22, 2009), pp. 1-13, ISSN 1741-7015.

Haldeman-Englert, C. & Jewett, T. (2011). 1q21.1 microdeletion, In: *GeneReviews,* R.A. Pagon et al. (Ed.), University of Washington, Seattle, Retrieved from http://www.ncbi.nlm.nih.gov/books/NBK1116/

Husi, H., et al. (2000). Proteomic analysis of NMDA receptor-adhesion protein signaling complexes. *Nature Neuroscience*, Vol. 3, No. 7 (July 2000), pp. 661-669, ISSN 1097-6256.

Kumar, A., et al. (2011). Animal model integration to AutDB, a genetic database for autism. *BMC Medical Genomics*, Vol. 4, pp. 15, ISSN 1755-8794.

Marshall, C.R., et al. (2008). Structural variation of chromosomes in autism spectrum disorder. *American Journal of Human Genetics*, Vol. 82, No. 2 (February 2008), pp. 477-488, ISSN 0002-9297.

Matuszek, G & Talabizadeh, Z. (2009). Autism Genetic Database: A comprehensive database for autism susceptibility gene-CNVs integrated with known noncoding RNAs and fragile sites. *BMC Medical Genetics*, Vol. 10, No. 102 (September 24, 2009), pp. 1-7, ISSN 1471-2350.

McClellan, J.M.; Susser, E. & King, M.-C. (2007). Schizophrenia: a common disease caused by multiple rare alleles. *British Journal of Psychiatry*, Vol. 190 (March 2007), pp. 194-199, ISSN 0007-1250.

McDonald-McGinn, D.M.; Emanuel, B.S. & Zackai, E.H. (2005). 22q11.2 deletion syndrome, In: *GeneReviews*, R.A. Pagon et al. (Ed.), University of Washington, Seattle, Retrieved from http://www.ncbi.nlm.nih.gov/books/NBK1116/

Miles, J.H., et al. (2010). Autism Spectrum Disorders, In: *GeneReviews*, R.A. Pagon et al. (Ed.), University of Washington, Seattle, Retrieved from http://www.ncbi.nlm.nih.gov/books/NBK1116/

Miller, D.T., et al. (2009). Microdeletion/duplication at 15q13.2q13.3 among individuals with features of autism and other neuropsychiatric disorders. *Journal of Medical Genetics*, Vol. 46, No. 4 (April 2009), pp. 242-248, ISSN 0022-2593.

Morrow, E.M., et al. (2008). Identifying autism loci and genes by tracing recent shared ancestry. *Science* Vol. 321, No. 5886 (July 11, 2008), pp. 218-223, ISSN 0036-8075

Nakatani, J., et al. (2009). Abnormal behavior in a chromosome-engineered mouse model for human 15q11-13 duplication seen in autism. *Cell*, Vol. 137, No. 7 (June 26, 2009), pp. 1235-1246, ISSN 0092-8674.

Noor, A., et al. (2010). Disruption at the PTCHD1 Locus on Xp22.11 in autism spectrum disorder and intellectual disability. *Science Translational Medicine*, Vol. 2, No. 49 (September 15, 2010), pp. 49-68, ISSN 1946-6234.

Phelan, K. (2007). 22q13.3 deletion syndrome, In: *GeneReviews*, R.A. Pagon et al. (Ed.), University of Washington, Seattle, Retrieved from http://www.ncbi.nlm.nih.gov/books/NBK1116/

Pinto, D., et al. (2010). Functional impact of global rare copy number variation in autism spectrum disorders. *Nature*, Vol. 466, No. 7304 (July 15, 2010) pp. 368-372, ISSN 0028-0836.

Rubenstein, J.L. & Merzenich, M.M. (2003). Model of autism: increased ratio of excitation/inhibition in key neural systems. *Genes, Brain & Behavior*, Vol. 2, No. 5 (October 2003), pp. 255-267, ISSN 1601-1848.

Schroer, R.J., et al. (1998). Autism and maternally derived aberrations of chromosome 15q. *American Journal of Medical Genetics*, Vol. 76, No. 4 (April 1, 1998), pp. 327-336, ISSN 0148-7299.

Sebat, J., et al. (2007). Strong association of *de novo* copy number mutations with autism.*Science*, Vol. 316, No. 5823 (April 20, 2007), pp. 445-449, ISSN 0036-8075.

Smith, C.L.; Bolton, A. & Nguyen, G. (2010). Genomic and epigenomic instability, fragile sites, schizophrenia and autism. *Current Genomics*, Vol. 11, No. 6 (September 2010), pp. 447-469, ISSN 1389-2029,

Szatmari, P., et al. (2007). Mapping autism risk loci using genetic linkage and chromosomal rearrangements. *Nature Genetics*, Vol. 39, No. 3 (March 2007), pp. 319-328, ISSN 1061-4036.

van der Zwaag, B., et al. (2009). Gene-network analysis identifies susceptibility genes related to glycobiology in autism. *PLoS One*, Vol. 4, No. 5 (May 28, 2009), pp. e5324, ISSN 1932-6203.

Vorstman, J.A.S., et al. (2010). A double hit implicates DIAPH3 as an autism risk gene. *Molecular Psychiatry* (published online March 23, 2010), ISSN 1359-4184.

Walsh, K.M. & Bracken, M.B. (2011). Copy number variation in the dosage-sensitive 16p11.2 interval accounts for only a small proportion of autism incidence: A systemic review and meta-analysis. *Genetics in Medicine* (published online February 1, 2011), ISSN 1098-3600.

Wang, K., et al. (2009). Common genetic variants on 5p14.1 associate with autism spectrum disorders. *Nature*, Vol. 459, No. 7246 (May 28, 2009), pp. 528-533, ISSN 0028-0836

Weiss, L.A., et al. (2009). A genome-wide linkage and association scan reveals novel loci for autism. *Nature*, Vol. 461, No. 7265 (October 8, 2009), pp. 802-808, ISSN 0028-0836.

Williams, C.A.; Dagli, A.I. & Driscoll, D.J. (2008). Angelman syndrome, In: *GeneReviews*, R.A. Pagon et al. (Ed.), University of Washington, Seattle, Retrieved from http://www.ncbi.nlm.nih.gov/books/NBK1116/

6

The Genetic Basis of Phenotypic Diversity: Autism as an Extreme Tail of a Complex Dimensional Trait

Shinji Ijichi[1,2], Naomi Ijichi[2], Yukina Ijichi[2], Hisami Sameshima[1] and Hirofumi Morioka[1]
[1]*Health Service Center, Kagoshima University, Kagoshima*
[2]*Institute for EGT (Externalization of Gifts and Talents), Kagoshima*
Japan

1. Introduction

Autism is a developmental lifelong condition of the human brain, and a behavioral characterization as a spectrum (autism spectrum disorder: ASD) is the best way to illustrate this complex trait (Frith, 2001; Rapin, 1997; Wing, 1997). The predominant presence of autistic cases without comorbidity (idiopathic or primary ASD) (Freitag, 2007) clearly means that the biological effects associated with the known concomitant medical conditions (cytogenic abnormalities, fragile X syndrome, tuberous sclerosis, congenital infections, maternal thalidomide use, epilepsy, etc.) cannot be the common prerequisite for ASD at least in the majority of the cases. The presence of a strong genetic contribution is evident from the results of twin studies, which demonstrated that 70-90% of monozygotic twins are concordant for ASD, and the concordance in dizygotic twins and the recurrence rate in the proband's siblings are both less than 10% (Rapin & Katzman, 1998). A broadening of the criteria of diagnosis leads the monozygotic concordance ratio to more than 90%, but 100% concordance is never obtained (Rapin & Katzman, 1998). Therefore, it is claimed that genetic factors contribute about 90% to ASD with environmental factors contributing no more than 10% (Garber, 2007). Although a flood of genetic information in the field of ASD is continuously growing, even the newest genome-wide molecular studies cannot detect the universal genetic prerequisite for idiopathic cases with ASD, compelling some researchers to speculate that ASD has a huge inter-case heterogeneity of the related gene variants.

Many gene variants, which seem to affect brain development and synaptic functions, have been reported in association with the autistic development (Betancur, 2011; Garber, 2007; Persico & Bourgeron, 2006; Pinto et al., 2010). In families with the candidates for autism gene variants, however, the strict co-segregation, in which the gene variant is found only in individuals with ASD among family members including parents, is still exceptional (Table 1). To explain this fact, the broader distribution of the more primary phenotype or pre-behavioral phenotype (endophenotype) beyond the categorical border is introduced as the speculative solution through this research maze (Viding & Blakemore, 2007). It may be quite difficult to detect and evaluate such endophenotypes because of the configurational or hierarchical structures of human cognitions and behaviors. Even if such speculations were all true, it is too early to conclude that "a single gene variant causes a small percentage of cases with this complex trait" (Garber, 2007; Beaudet, 2007). As clearly demonstrated in the case of human disease-associated mutations found as wild-type alleles in normal chimpanzee (L. Azevedo et al., 2006), a deleted or mutated allele does not necessarily contribute to the disease development. Because evidence consistent with a theory is not proof of that theory (Cannell, 2010), until one could delineate the molecular or biological trajectory underlying autistic development which is quantitatively different from the parents, there is still a huge black box between the *de novo* variant allele and complex human behaviors in the sporadic cases with idiopathic ASD. The reported gene variants are, at present, nothing but one of the concomitants in a small percentage of cases (5-7%, in Table 1). The possibility that the variantsare mere relative risk factors remains to be elucidated (Jones & Szatmari, 2002). As a general rule, a genetic link does not necessarily imply neurological damage (Simpson,

Variants	Prevalence		References
SHANK3 variants	In ASD families	15 / 227 (6.6%)[a]	(Durand et al., 2007)
	Co-segregated cases	3 / 227 (1.3%)[b]	
	In controls	5 / 270 (1.9%)[a]	
SHANK3 variants	In ASD families	3 / 400 (0.8%)[c]	(Moessner et al., 2007)
SHANK3 variants	In ASD individuals	34 / 427 (8.0%)	(Gauthier et al., 2009)
	In controls	16 / 190 (8.4%)	
SHANK3 deletion	In ASD individuals	1 / 427 (0.2%)[c]	(Gauthier et al., 2009)
	In controls	0 / 190	
SHANK3 deletion	In ASD individuals	2 / 2,195 (0.1%)	(Glessner et al., 2009)
	In controls	2 / 2,519 (0.1%)	
SHANK2 *de novo* deletion	In ASD individuals	2 / 996 (0.2%)	(Pinto et al., 2010)
	In controls	0 / 1,287, 0 / 3,677	
NLGN3 variants	In ASD individuals	0 / 96	(Yan et al., 2005)
NLGN3 duplication	In ASD individuals	1 / 2,195 (0.05%)	(Glessner et al., 2009)
	In controls	0 / 2,519	
NLGN4 variants	In ASD individuals	4 / 148 (2.7%)[d]	(Yan et al., 2005)
	In controls	0 / 336	
NRXN1 deletion	In ASD families	1 / 1,181 (0.1%)	(AGPC, 2007)
NRXN1α variants	In ASD individuals	5 / 116 (4.3%)	(Yan et al., 2008)
	In controls	1 / 192 (0.5%)	
NRXN1 deletion	In ASD individuals	10 / 2,195 (0.5%)	(Glessner et al., 2009)
	In controls	0 / 2,519	
NRXN1 *de novo* variants	In ASD families	4 / 996 (0.4%)	(Pinto et al., 2010)
	In controls	5 / 1,287 (0.4%)	
NRXN1β variants	In ASD individuals	4 / 203 (2.0%)[d,e]	(Feng et al., 2006)
	In controls	0 / 535	
NRXN2β variants	In ASD individuals	0 / 72	(Feng et al., 2006)
NRXN3β variants	In ASD individuals	0 / 72	(Feng et al., 2006)
CNTN4 deletion	In ASD individuals	10 / 2,195 (0.5%)	(Glessner et al., 2009)
	In controls	0 / 2,519	
CNTN4 duplication	In ASD individuals	9 / 2,195 (0.4%)[c]	(Glessner et al., 2009)
	In controls	1 / 2,519 (0.04%)	

Variants	Prevalence		References
AUTS2	In ASD individuals	1 / 2,195 (0.05%)	(Glessner et al., 2009)
	In controls	0 / 2,519	
DDX53/PTCHD1 deletion (maternally inherited)	In ASD cases	7 / 996 (0.7%)	(Pinto et al., 2010)
	In controls	0 / 1,287, 0 / 3,677	
CNVs at 15q11-13 (UBE3A)	In ASD individuals	15 / 2,195 (0.7%)[f]	(Glessner et al., 2009)
	In controls	0 / 2,519	
CNVs at 15q11-13	In ASD individuals	4 / 522 (0.8%)[g]	(Depienne et al., 2009)
CNVs at 16p11.2	In ASD families	12 / 751 (1.6%)	(Weiss et al., 2008)
	In controls	5 / 4,234 (0.1%)	
	In ASD individuals	3 / 299 (1.0%)	
	In controls	7 / 18,834 (0.04%)	
16p11.2 duplication	In ASD individuals	9 / 2,195 (0.4%)[c]	(Glessner et al., 2009)
	In controls	4 / 2,519 (0.2%)	
16p11.2 deletion	In ASD individuals	8 / 2,195 (0.4%)[h]	(Glessner et al., 2009)
	In controls	4 / 2,519 (0.2%)	
CNV gain at 1q21	In ASD families	3 / 1,181 (0.3%)[d]	(AGPC, 2007)
CNV at 17p12	In ASD families	3 / 1,181 (0.3%)[i]	(AGPC, 2007)
CNV gain at 22q11.2	In ASD families	2 / 1,181 (0.2%)[d]	(AGPC, 2007)
22q11.2 duplication	In ASD individuals	9 / 2,195 (0.4%)	(Glessner et al., 2009)
	In controls	0 / 2,519	
De novo CNVs	In ASD families	10 / 1,181 (0.8%)	(AGPC, 2007)
	Co-segregated cases	3 / 1,181 (0.3%)[j]	
De novo CNVs	In ASD individuals	14 / 195 (7.2%)[k]	(Sebat et al., 2007)
	In sporadic cases	12 / 118 (10.2%)	
	In multiplex families	2 / 77 (2.6%)[k]	
	In controls	2 / 196 (1.0%)	
De novo CNVs	In ASD families	27 / 427 (6.3%)	(Marchall et al., 2008)
	In sporadic cases	4 / 56 (7.1%)	
	In multiplex families	1 / 49 (2.0%)	
De novo CNVs	In ASD families	50 / 876 (5.7%)[l]	(Pinto et al., 2010)
	In simplex families	22 / 393 (5.6%)	
	In multiplex families	19 / 348 (5.5%)	

ASDs: autism spectrum disorders; NLGN: neuroligin gene; NRXN: neurexin gene; CNTN: contactin gene; AUTS: autism susceptibility candidate gene; CNV: copy number variation; AGPC: the Autism Genome Project Consortium. [a]Two nonsynonymous SHANK3 mutations were revealed in 4 ASD families and 2 control individuals. [b]In the SHANK3 study, *de novo* truncating mutations in two families and a chromosomal rearrangement in one family were demonstrated as the strict co-segregated cases whose gene variants were found only in individuals with ASD among family members including parents. [c]One *de novo* case is included. [d]Strict co-segregation was not shown. [e]Two cases with mild facial dysmorphism are included. [f]Two *de novo* cases are included. [g]Three *de novo* cases are included. [h]Five *de novo* cases are included. [i]One case is included as a co-segregated family. [j]In ASD families with two or more affected individuals (multiplex families), three *de novo* CNVs were found in both ASD sibs. [k]Two multiplex families whose variant-phenotype co-segregation is not mentioned are included. [l]>0.6% cases are carrying two or more *de novo* events.

Table 1. The prevalence of variants in gene regions recently implicated in idiopathic ASD

2003), and high heritability does not vindicate the condition as a diagnostic category (Keller & Miller, 2006). There is as yet no qualitative biological marker including microscopic lesions that can reliably help to categorize a genetically homogeneous autism subtype (Amaral et al., 2008; Moldin et al., 2006; Santangelo & Tsatsanis, 2005; Schmitz & Rezaie, 2008). In this article, the significance of gene variants which have currently been detected in autistic individuals is carefully reconsidered and the outstanding questions are addressed from multidisciplinary points of view. Such an attempt may highlight the importance of the notion that the evolutionally survived trait is the phenotypic diversity itself, in which ASD is included as an extreme tail. In addition, important concepts and mechanisms for the genetic basis of phenotypic diversity are also reviewed.

2. Facts and questions

Although some authorities appreciated the smooth behavioral continuum between individuals with ASD and the non-autistic majority (Frith, 2001; Happé, 1999; Rapin, 1997; Wing, 1997), idiopathic ASD has sometimes been misinterpreted as a qualitative disorder which can be clearly distinguished from normal development. The boundary between individuals with low-functioning ASD and a communicative subtype (Asperger syndrome) has also been misrepresented as to be qualitatively distinct (Simpson, 2003). Even the differentiation between Asperger syndrome and high-functioning ASD could be made with authority (Kamp-Becker et al., 2010). These biased constructions may be attributable to referral bias in general practice or increased probability of clinical ascertainment in individuals with low achievement (Skuse, 2007). Although ASD can still be documented as a categorical entity in clinically ascertained samples (Frazier et al., 2010), the fact that the autistic phenotype extends beyond its formal diagnostic boundaries has underscored the significance of quantitative evaluations (Lamb et al., 2000; Maestrini et al., 2000), and many population studies revealed that ASD including high-functioning subtypes are best characterized as an extreme of some bell-shaped behavioral dimensions that distribute quantitatively (Constantino & Todd, 2000; Constantino & Todd, 2003; Happé et al., 2006; Hoekstra et al., 2007; Posserud et al., 2006; Ronald et al., 2005; Ronald et al., 2006a, 2006b; Skuse et al., 2005). The description 'qualitative' in the autism criteria in the Diagnostic and Statistical Manual of Mental Disorders (DSM) is removed and Asperger's disorder (Asperger syndrome) is subsumed into ASD in the draft of DSM-5 (http://www.dsm5.org /Pages/Default.aspx). The three quantitative domains including sociability, communication, and rigid/repetitive behavior correlate modestly to each other in the population (Dworzynski et al., 2007; Ronald et al., 2005, 2006a, 2006b), and the coincidence of these phenotypic extremes is also observed in hyperactive individuals with attention-deficit/hyperactivity disorder (AD/HD) (Hattori et al., 2006; Ijichi & Ijichi, 2007; Reiersen et al., 2007; Ronald et al., 2008). The diagnosis of autism is highly affected by the circumstantial consequence of social adaptability and autistic recognition and behavior sometimes does not become fully manifest until social demands exceed the individual's limited capacities (the draft of DSM-5). The clinical picture can change with increasing age and in different circumstances (Wing, 1997), and the behavioral plasticity or clinical improvement is evident in supportive circumstances by structured behavioral interventions, mentoring, and/or social involvement with appropriate accommodation (Garcia-Villamisar & Hughes, 2007; Ijichi & Ijichi, 2007; McGovern & Sigman, 2005; Tonge et al., 1994).

The most unique and potentially meaningful property of autistic cognition is savant skill. The estimated prevalence of the cognitive superiority in ASD varies from 10% to surprising numbers (Dawson et al., 2007; Happé, 1999; Rapin & Katzman, 1998). The supposed common 'high intelligence' in autistic individuals with low IQ may involve high processing speed, prodigious memory capacities, and heightened primary sensory processing (Boddaert et al., 2005; McCleery et al., 2007; Scheuffgen et al., 2000). These cognitive superiorities are believed to have the same origin as the social difficulties in ASD (Brosius & Kreitman), and the term, 'autistic savant skills', is used to describe one of the core cognitive features of ASD (Badcock & Crespi, 2006; Scheuffgen et al., 2000). As a unifying explanation which covers the manifold autistic characteristics, excessive neuronal processing (a hyper-functionality model) is also implicated as opposed to usual hypo-functionality explanations (Markram et al., 2007).

The ratio of sibling recurrence risk to population prevalence is approximately 50 with an overwhelming predominance of sporadic cases, suggesting the multifactorial nature of ASD (AGPC, 2007). The high monozygotic concordance rate in twins and the modest recurrence risk in dizygotic twins and among siblings may also suggest that the genetic architecture for ASD has the same complexity as those for human physical appearances including facial characteristics and brain gray matter volume (Ijichi & Ijichi, 2004). In traditional views, the modest correlation between autistic behavioral domains in population studies implies that there is no single (genetic or endophenotypical) cause for the three autistic extreme characteristics and a mere coincidence of the phenotypic extremes might be the true nature of autistic social difficulties (Happé et al., 2006). Although positive assortative mating might cause phenotypic anticipation and a negative assortative mating between the couple might gather the non-overlapping genetic components in a baby (Ijichi et al., 2008), there is no evidence for such assortative mating (Hoekstra et al., 2007).

As exemplified in Table 1, there is, so far, no universal genetic marker which is co-segregated with ASD in the affected families. In contrast to the early prediction (30-40%) (Beaudet, 2007), no more than 5-7% of ASD cases may be traceable to single or multiple genetic concomitant(s) (Table 1). Although many whole-genome scans for autism susceptibility loci have identified a lot of linkage peaks, the reproduction of the results is exceptional and association studies have failed to identify the gene variants (Sykes & Lamb, 2007). The regions of structural variants including copy number variations (CNVs) seldom conform to the linkage peaks (Sebat, 2007). The lack of an unambiguous pathophysiological marker is also one of the important characteristics of idiopathic autism (Amaral et al., 2008; Moldin et al., 2006; Santangelo & Tsatsanis, 2005; Schmitz & Rezaie, 2008). The only anatomical candidate which can be consistently co-segregated with ASD including masked autistic savants may be a quantitative increase in the number of processing units of cortex (minicolumns) (Casanova, et al., 2002, 2007). The increase in the number of minicolumns is thought to be associated with mammalian brain evolution, and the finding can explain other apparent tendencies revealed in some autistic individuals, including increases in the volume of brain structures and the prevalence of epilepsy (Casanova et al., 2006). Recent preliminary findings suggest that the tendency of brain overgrowth originates prenatally (Hobbs et al., 2007; Leonard et al., 2008). Furthermore, there is no biological deficit including chemical and molecular findings which is universal in individuals with ASD or can reliably help to identify putative subgroups that are genetically homogeneous (Lauritsen & Ewald, 2001). Over-expression of neuron-associated genes is still one of the candidates for molecular markers (Lepagnol-Bestel et al., 2008; Maussion et al., 2008; Rinaldi et al., 2007).

The scientific puzzle, which is metaphorically described as "myopic investigators are still patting the elephant" (Rapin, 1999) remains to be solved (Baron, 2008). Why is the male to female ratio biased (3-4 to 1)? Why cannot the behavioral uniformity with strong genetic contribution be interpreted by common gene variant alleles? Why is the disparity between monozygotic and dizygotic concordances so large? Why do the autistic behavioral domains correlate modestly? Although these questions may provide very important clues and are encouraging researchers to speculate on reasons, the jigsaw is still incomplete. Missing puzzle pieces include the solution of the evolutionary mystery of autism prevalence. Human conditions can be selected and survive when it is somehow associated with increased reproductive success (Nesse & Williams, 1994). However, in spite of the hypo-reproductive tendency of behaviors in extreme cases with ASD (Lord et al., 2000), the estimated high prevalence has never declined (Baird et al., 2006; CDC, 2009; Fombonne, 2009).

3. Genetic and environmental explanations

It is recently recognized that ASD has the highest prevalence (more than 0.5%) in childhood neurodevelopmental conditions (CDC, 2009; Fombonne, 2009). In traditional frameworks, in which researchers are searching the human genome for the condition-specific genetic variants, three genetic models should be considered as the genetic mechanism for such a common phenotypic condition (Gibson, 2009). The quite low effect size of each ASD-related variant is suggested to be the cause of difficulty in replication of the positive findings in the common disease-common variant (CD-CV) model (Anney, 2010). Although a rare alleles of major effect (RAME) model is one of the core principles for recent genome-wide association studies in ASD (Gibson, 2009), the replication may also be complicated by chance findings, as well as differences in ascertainment, because of the modest relative risk of the rare alleles (Anney, 2010). The third model, the infinitesimal model, can make an excuse for the situation of genetic studies, because it is very hard to identify rare variants of small effect by genetic means (Manolio et al., 2009). It is, anyhow, clear that it's time to reconsider and question simple intuitive models that link a human complex condition to mutation (Gibson, 2009).

3.1 Genetic factors

The non-universality of the candidate gene variants which have previously been implicated in ASD may be consistent with the speculation that heterogeneous sets of gene variants can contribute to ASD (Betancur, 2011; Beaudet, 2007; Garber, 2007). Furthermore, in order to explain the modest correlations between the three autistic behavioral domains, the presence of domain-specific heterogeneous sets of gene variants are also suggested (Happé et al., 2006). However, even novel genetic means including whole genome screening using microarray-based hybridization cannot fully confirm these speculations (Table 1). The frequent absence of diagnostic history of ASD in the parents of an idiopathic ASD proband may suggest that the supposed variants should be carried by a non-ASD parent (incomplete penetrance) or the proband should have *de novo* mutations (Beaudet, 2007; Constantino & Todd, 2005; Zhao et al., 2007) (Table 2). However again, such genetic transmission is still one of the hypotheses and the concomitant *de novo* variants can be detected only in a minor part of the cases (Table 1). The number of candidate gene regions is still increasing without a convincing and comprehensive demonstration of the link between such variants and autistic developmental trajectory (Glessner et al., 2009).

The genetic contribution to a quantitative trait may be attributable to the cumulative effect of a set of quantitative trait loci (QTLs) (Plomin et al., 1994, 2009; Plomin & Kosslyn, 2001). Each QTL is neither necessary nor sufficient for the overall phenotypic outcomes, the effect size of each QTL may fluctuate according to other genetic backgrounds (epistasis, non-additive gene-gene interactions) and the environment (gene-environment interactions), and a QTL may affect more than one phenotypic trait (pleiotropy). The concept of epistasis had initially been introduced for ASD as an alternative explanation of the incomplete penetrance or as a risk factor model (Bradford et al., 2001; Folstein & Rosen-Sheidley, 2001; Jones & Szatmari, 2002). Because natural chromosomal and segmental shuffling during normal meiosis is a strong random modifier of epstatic effects among QTLs in a sib-pair and dizygotic twins, the big disparity between monozygotic and dizygotic concordances in autism may be explained by the presence of epistatic QTLs. Pleiotropy can account for the presence of autistic savants. The modest correlation among autistic behavioral domains can also illustrated by unsynchronized epistatic pleiotropy (Ijichi et al., 2008). To explain the sporadic manner of the prevalence and the survival of hypo-reproductive autistic extremes, the implication of epistasis-associated intergenerational oscillation of phenotypic outcomes was introduced (Ijichi et al., 2008). Some candidates for autism QTLs have been reported (Ashley-Koch et al., 2006; Coutinho et al., 2007; Jiang et al., 2004; Weiss et al., 2007), linkage analysis with quantitative measures of some autistic characteristics revealed QTL signals (Alarcón et al., 2002, 2005; Chen et al., 2006; Duvall et al., 2007), and a quantitative covariance analysis can confirm the high genetic correlation between 'social motivation' and 'range of interest/flexibility' (Sung et al., 2005). Although the supposed contribution of QTLs ought to be traced in family studies or genome scans according to a traditional logic, "the causal gene variant can be cosegregated with the phenotypic variant", the delay and difficulty in detecting the causal variant alleles at QTLs is strangely common to all idiopathic quantitative traits including autism, physical and physiological characteristics, and personalities (de Geus et al., 2001; Fullerton, 2006; Palmert & Hirschhorn, 2003; Willis-Owen & Flint, 2006).

Facts and questions	Explanations			
	Penetrance	*De novo*	QTLs	Environment
The quantitative feature	(○)	—	○	(○)
Partial behavioral plasticity	—	—	—	○
The presence of autistic savants	—	—	(○)	—
Strong genetic contribution	○	○	○	—
Usually sporadic without family history	(○)	○	(○)	○
Domain-specific genetic factors	—	—	(○)	—
Lack of the common genetic marker	○	○	○	—
Lack of the common pathological lesion	—	—	—	—
Lack of the common chemical marker	—	—	—	—
Lack of the common molecular marker	—	—	—	—
Why is the male to female ratio biased?	(○)	(○)	(○)	(○)
Why is it so difficult to detect autism genes?	—	—	—	—
Why do hypo-reproductive extremes survive?	—	○	—	(○)

Penetrance: Poor penetrance of heterogeneous gene variants; *De novo*: *De novo* involvement of heterogeneous gene variants; QTLs: Quantitative trait loci; Environment: Environmental contribution; ○: explainable; (○): unexplainable by itself but explainable with some further speculation; —: hard to explain

Table 2. Genetic and environmental explanations for the facts and outstanding questions in idiopathic autism researches

3.2 Epigenetic factors and ASD

Phenotypic outcomes with robustness or plasticity cannot be exclusively determined by the DNA sequence itself which looks like the core genetic factor of the phenotype (Goldberg et al., 2007). Epigenetics is the study of changes in gene expression that occur without a change in DNA sequence and the epigenotype is meiotically and mitotically transmissible (Morris, 2005; van Vliet et al., 2007). Although the significance of the contribution made by epigenetic factors to human complex traits remains unclear, it is speculated that epigenetic factors can influence gene-environment interactions and the liability/outcomes of the traits (van Vliet et al., 2007). Epigenetic changes in gene expression are achieved through RNA-associated silencing, DNA methylation, and histone modifications (Morris, 2005), and cis-acting expansion of the epigenetic influences on the flanking genes is referred to as genomic imprinting which results in parent of origin-specific gene expression (Pauler et al., 2007). The epigenetic factors and genomic imprinting may be implicated in syndromic autistic individuals with some single gene/chromosomal disorders including Rett syndrome, fragile X syndrome, Prader-Willi syndrome, and Angelman syndrome, and a variety of factors associated with epigenetic modifications have been considered as candidates for autism genes (Badcock & Crespi, 2006; Jiang et al., 2004; Persico & Bourgeron, 2006; Schanen, 2006; Skuse, 2000; van Vliet et al., 2007). These factors, however, cannot be the common prerequisite for idiopathic ASD at least in the majority of the cases (Jiang et al., 2004; Persico & Bourgeron, 2006), and the power of epigenetic factors are recognized as an accidental cue to shift the quantitative distribution of the autistic traits in a threshold model (Skuse, 2000). If the epigenetic factors act only in gene-environment interactions in idiopathic cases, the epigenetic contribution should be modest in the overall underpinnings. Given an unforeseen transmissible powerful architecture connecting genotype and phenotype for phenotypic diversity independent of genetic diversity, the epigenetic mechanism should be referred to as merely one of the molecular-level environments derived from gene networks.

3.3 Environmental factors

Environmental factors contribute no more than 10% to ASD (Garber, 2007). However, the environmental factors including rubella, thalidomide, and valproic acid embryopathies may still be important as additive triggers of the clinical manifestation (Jones & Szatmari, 2002; Persico & Bourgeron, 2006). Environmental contributions including behavioral experiences are originally misunderstood to explain the patterns of familial recurrence risks observed in autism studies (Jorde et al., 1991). Because the genetic components affecting autistic traits seem to be the same across the sexes (Constantino & Todd; Hoekstra et al., 2007), it can be speculated that the lower prevalence of autistic traits in girls is the result of increased sensitivity to early environmental influences that operate to promote social competency (Constantino & Todd, 2003). The minimal contribution of shared environmental influences (Ronald et al., 2006a) may be associated with the autistic behavioral manifestations including resistance to change or insistence on sameness.

Combinations of the traditional theories (poor penetrance, *de novo* mutations, and QTLs and the environmental contribution) may answer not a few of the outstanding questions in idiopathic autism research (Table 2). However, in spite of the presence of a big genetic contribution to the autistic development, the question, "Why is it difficult to detect autism gene variants?", still remains to be resolved. In addition, the significance of both *de novo* mutations and the environmental modification is just a speculation in a part of the ASD cases.

4. Evolutionary explanations

Does idiopathic ASD really represent many distinct conditions with numerous etiologies (Geschwind, 2007)? Is it really time to give up on a single explanation for autism (Baron, 2008; Happé et al., 2006)? A variety of qualitative concomitants, including gene variants and environmental factors, have already been demonstrated in part of autistic cases as exemplified above. However, it may be still too early to reach the conclusion even in such frameworks, because no single qualitative process associated with the concomitants can indicate the molecular or chemical differences between the autistic developmental extremes and the non-autistic majority. In order to understand human complex traits, genetic, molecular, and biochemical explanations should be combined with evolutionary explanations (Nesse & Williams, 1994). In autistic individuals, ASD *per se* does not shorten the span of life (Gillberg et al., 2010). Although high or preserved androgenic competence is suspected in ASD (Tordjman et al., 1997), the extreme cases almost never marry (Lord et al., 2000). The hypo-fertility results from reduced opportunity or behavioral ability in the mating arena. Therefore, we must probe into who is enjoying the reproductive benefits of the genetic architecture for ASD in the evolutionary framework (Table 3).

Who gets the reproductive benefits?	Hypotheses or mechanisms	References
None (an inevitable outcome)	Mutation-selection balance theory	(Keller & Miller, 2006)
Unaffected carriers of genetic factors	Hyper-systemizing theory (extreme male brain theory) Extreme imprinted brain theory	(Baron-Cohen, 2002) (Badcock & Crespi, 2006)
All of the non-autistic majority	Population benefit theory Monomorphic loci theory	(Fitzgerald, 2002) (Ijichi et al., 2011)

Table 3. Evolutionary explanations for the survival of autistic extremeness

4.1 Mutation-selection balance theory

In the mutation-selection balance theory, individuals with a high load of mutations are postulated to be at higher chance of passing risk on to their offspring, and it is not necessary that there are individuals with the reproductive benefits (Keller & Miller, 2006). Importantly, according to the proposed model, everyone alive has minor brain deviations that cause them to be a little bit abnormal in behavioral and cognitive dimensions (Keller & Miller, 2006). The non-autistic majority in the population is regarded as the genetic carrier-state for ASD and the mutation load and the risk of having autistic offspring may vary quantitatively. In the mutation-selection balance theory, balancing selection for genetic diversity is recognized to be unsuitable to explain persistent heritability in human conditions (Keller & Miller, 2006; Zhang & Hill, 2005). One of the grounds of this exclusion of balancing selection is the absence of an ongoing homeostatic mechanism that counteracts the homogenizing effect of genetic drift and stabilizing selection, and the reproductive benefits of the genetic burden for autism are not addressed (Keller & Miller, 2006). The mutation-selection perspective can be an evolutionary interpretation of a cumulative effect of *de novo* mutations and is at least consistent with the quantitative distribution of autistic domains.

4.2 Extreme male brain theory

The second is a group of theories in which only a part of the population is regarded as the genetic carrier-state for ASD. The prevalence or maintenance of positive assortative mating between the non-autistic carriers is critical to accumulate genetic factors in these theories, and the remaining non-autistic majority does not have the genetic components for ASD. In the hyper-systemizing theory, the unaffected carriers of the genetic factors are high systemizers and ASD is the result of both parents being the high systemizers (Baron-Cohen, 2002, 2004, 2006). Systemizing is the drive to understand and predict the next step of inanimate events and acts contrary to empathizing. In males, the systemizing mechanism is set at a slightly higher level than non-autistic males (Baron-Cohen, 2004). This extreme male brain theory of autism had originally been proposed by Asperger in 1944. Individuals including both parents of individuals with autism, who are placed in the adjacent part to the autistic extremeness, systemize at a higher level than average (above average systemizers) and account for approximately a half of the vast majority. Over successive generations, the above average systemizers carry the genetic components for ASD and might enjoy the reproductive benefits. As one of the genetic bases of the hyper-systemizing theory, the extreme imprinted brain theory had been proposed (Badcock & Crespi, 2006).

4.3 Population benefit theory and individual benefit theory

In the third framework, it is suggested that the evolutionarilly selected and conserved phenotype is not the hypo-reproductive extremeness but the whole quantitative distribution itself. A group selection theory has been introduced to bring sense into the link between autism and exceptional creativity (Fitzgerald, 2003). In this population benefit theory, the creativity, which can be concomitant with autism, benefits all members of the human community and the community can survive. On the other hand, the third framework can also include individual benefit concepts (the monomorphic loci theory) (Ijichi et al., 2011). In the individual benefit concepts, everybody has both the genetic architecture for ASD and the possibility to enjoy the reproductive benefits of autism genes. Each phenotypic outcome, however, varies individually mainly according to the differences in genetic background noise and environmental factors, whose functions are not necessarily related to ASD phenotypes directly. In the process of reaching the monomorphic loci theory, the epistasis-mediated intergenerational oscillation of phenotypic outcomes has been advanced in a QTL model (Ijichi et al., 2008). The monomorphic loci theory does not dismiss the comprehensive view of the known genetic contributions, including major gene effects and additive genetic networks (Ijichi et al., 2011). The postulated involvement of monomorphic loci can be valid as merely one of the genetic constituents in complex (additive and/or non-additive) interactions with polymorphic loci.

4.4 The monomorphic loci theory and gene networks

Because both positive and negative epistasis may be byproducts of evolution (L. Azevedo et al., 2006; R.B.R. Azevedo et al., 2006; Harrison et al., 2007), the invisibility of the contribution of monomorphic epistatic loci from the traditional genetic view is an attractive candidate for the explanation of the black box between polymorphic genotype and phenotypic diversity (Ijichi et al., 2011). Complex phenotypes have hierarchical structures, including RNA (transcript traits), protein, metabolite, and functional levels. It has been suggested that less heritability of metabolite traits than transcript traits is associated with the difference in the quantity of biological noise between the genetic determinants and the trait (Rowe et al., 2008). The more steps that are involved between genotype and the trait level, the more biological noise may reside in the process. Such biological noise originates from inter-locus interactions and gene-environment interactions, and the inter-locus interactions may have an important role in the biological noise. Additive and/or non-additive inter-locus interactions with other loci are available in a variety of processes including cis-, trans-, and inter-cellular interactions (Figure 1). The presence of gene-environment-gene circuits may make it difficult to distinguish inter-locus interactions from gene-environment interactions in the biological noise (Ijichi et al., 2011). In these interactions, an intergenerational change in the number or property of factors (environment and/or other related loci) in the regulatory circuit may easily individualize the balance of each hierarchical trajectory (coding RNA, non-coding RNA, translation, autocrine, paracrine, and endocrine levels) and individually determine the developmental outcomes. The net non-additive effects of the biological noise are metaphorically interpreted as hub-and-spoke structures of regulatory networks among polymorphic loci (Benfey & Mitchell-Olds, 2008).

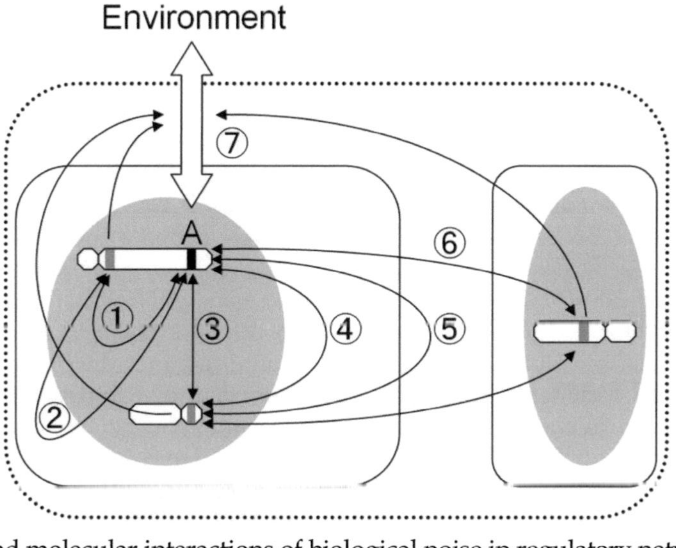

Fig. 1. Cellular and molecular interactions of biological noise in regulatory networks around a gene locus (A). Additive and/or non-additive phenomena can be involved in each interaction (Ijichi et al., 2011). In this explanation, an arrow represents the net contribution between loci and the gene-environment relationship. The locus A can interact with other loci in association with coding RNA and/or non-coding RNA level in cis-acting manner (①, ②) and trans-acting manner (③, ④). The cis-acting interactions are involved in genetic imprinting. After translation, interactions can be mediated through autocrine, paracrine, and endocrine mechanisms (⑤, ⑥). Gene-environment interactions can modify penetrance of the outcomes affected by the locus A. The network constituents can change the sensitivity to environmental influences (⑦), that can provide gene-environment-gene circuits. In the monomorphic loci theory, the gene A can be monomorphic and the link between monomorphic A and the A-associated polymorphic noise is usually invisible in the context of traditional genetics.

5. Quantitative domains and genetic factors

The distributional shift of a bell-shaped curve and the change in the curve shape illustrates the mean value change and the variance alteration of the quantitative dimension, respectively (Gibson, 2009). These changes can affect the proportion of individuals with autism to those without as determined by a liability threshold. The biased male to female ratio (3-4 to 1) in ASD is plausibly interpreted as a distributional shift of the quantitative bell-shaped curve as a gender gap. In the hyper-systemizing theory, the male systemizing mechanism is set at a slightly higher level than in females (Baron-Cohen, 2004). In an imprinted-X liability threshold model, actions of some X-linked genes, which are expressed only from paternal X-chromosome, are suggested to be associated with the male predisposition to ASD (Skuse, 2000). The gender is a bimorphic genetic variation and there is a gender gap in sensitivity or vulnerability to environmental factors (Constantino & Todd, 2003). The relationship between a bell-shaped quantitative distribution and the genetic factors underlying the complex phenotype still remains to be elucidated.

5.1 Polygenic liability model

The traditional concept of polygenic liability supposes a normal distribution of frequencies of susceptibility variant alleles (Gibson, 2009). The manner of the allele contribution is additive, and each allele contribution usually results in a positive or negative effect on the phenotype in the carrier individual and the quantitative population dimension results from such additive allele contributions. To explain the smooth normal distribution, an environmental variance of each allele contribution is addressed in this model.

In a genetic model, oligogenicity with epistasis, the contributing genes are likely to be common ones in the population (Folstein & Rosen-Sheidley, 2001). There is no evidence that the genetic causative processes affecting the autistic extreme are different from those contributing the autistic dimension including individuals without autism (Ronald et al., 2006a). If the presence of epistasis, pleiotropy, and gene-environment interactions are all supposed, the polymorphic genetic underpinning is referred to as QTLs (Plomin et al., 1994, 2009; Plomin & Kosslyn, 2001). However, it is also the fact that the delay and difficulty in detecting the causal variant alleles at QTLs is common to all idiopathic quantitative traits including ASD, physical and physiological characteristics, and personalities (de Geus et al., 2001; Fullerton, 2006; Palmert & Hirschhorn, 2003; Willis-Owen & Flint, 2006).

If the genetic factors for a tail of the bell-shaped curve are different from those for the majority and have extremity-specific properties including serious involvement of coding gene segments (Mitchison, 2000), the variant alleles should be more detectable. Because the genetic contribution in ASD is the biggest in human complex traits and the environmental influence on ASD is quite minimal as described above, the difficulty in finding the universal genetic marker for ASD warrants the necessity of a paradigm shift.

5.2 Additive and non-additive interactions between mono- and poly-morphic loci

It has been emphasized that the three behavioral domains of ASD modestly correlate to each other and the set of genes for each domain may be partly different (Dworzynski et al., 2007; Happé et al., 2006; Ronald et al., 2005, 2006a, 2006b). The speculated modest genetic overlap among autistic domains may be indistinguishable from that among human complex phenotypes including ASD, bipolar disorder, and schizophrenia (Rzhetsky et al., 2007), suggesting that the autistic domains and these psychiatric conditions might share the same genetic architecture at least in part (Craddock & Owen, 2010). In an argument about domain-specific genes for cognitive functions, it is expected that the domain-general genes are responsible for the brain infrastructure including receptors, neurotransmitters, dendritic spines, synapse vesicles, and axonal filaments (Marcus & Rabagliati, 2006). Although the universality of the domain-general genes for cognitive functions among other human complex phenotypes is controversial, genes for the brain infrastructure are also current topics in the field of ASD (Garber, 2007; Persico & Bourgeron, 2006). Both the heterogeneity of genetic markers for ASD and the modest correlation among autistic core domains can be explained by epistasis-mediated oscillation of the domain-general effect values and unsynchronized epistatic pleiotropy in the monomorphic loci theory, which never dismiss the comprehensive view of the known genetic contributions, including major gene effects and additive genetic networks (Ijichi et al., 2011). The assumption of the random outcomes mediated by the non-additive interactions between functional monomorphic loci and polymorphic backgrounds may transform the traditional complementary roles of some monomorphic loci (Gjuvsland et al., 2007) to active and leading roles for the phenotypic diversity (Ijichi et al., 2011). However, the controversy concerning the importance of non-additive effects in phenotypic diversity still exists (Gale et al., 2009; Hill et al., 2008; Malmberg & Mauricio, 2005).

5.3 Social environmental changes and decanalization

The decanalization concept may have sizable significance in searching the cause of the maintained or increasing prevalence of ASD. Canalization is an evolutionary phenomenon characterized by robustness to genetic or environmental perturbation, and most individuals tend to cluster around the optimal phenotype in canalized populations (Gibson, 2009). If the phenotypic dimension consists of multiple endophenotypic vectors which have nonlinear relationships to each other and are partially determined by genetic factors, overt environmental perturbations for one of the endophenotypes can be the cue of decanalization, which changes the shape of the phenotypic demensional distribution (Gibson, 2009). Social environmental perturbations may also shift the entire distribution of ASD liability, or move the liability threshold.

6. Conclusions

The difficulty in detecting the universal biological marker for the predisposition to ASD presents significant challenges and conflicts to researchers in related fields. The reported gene variants in some sporadic cases with idiopathic ASD are nothing but one of the concomitants, until the molecular or biological trajectory underlying autistic development is clearly delineated or association studies reproduce the causal relationship. Before the speculation that idiopathic ASD represents many distinct conditions with numerous etiologies, the quantitative manner of the distribution of the behavioral domains and the fact that ASD is a mere tail of the behavioral dimensions should strictly be considered and emphasized. Even combinations of traditional theories including poor penetrance, *de novo* mutations, quantitative trait loci, and environmental contribution cannot fully account for the entire genetic underpinning. Importantly, the almost monolithic insight into the prevalence of ASD can only be obtained in an evolutionary framework on the assumption that the complex genetic networks are responsible not for the individual cases but for the human behavioral diversity itself. Gender differences, environmental factors, epigenetic mechanisms including genetic imprinting, and major gene effects may all be mere accidental modifiers of the relationship between the diversity and the liability threshold.

7. Acknoledgment

The authors greatly thank Professor Bryan H. King for his helpful suggestions.

8. References

Alarcón, M., Cantor, R.M., Liu, J., Gilliam, T.C., the Autism Genetic Resource Exchange Consortium & Geschwind, D.H. (2002). Evidence for a language quantitative trait locus on chromosome 7q in multiplex autism families. *Am J Hum Genet*, 70, 60-71

Alarcón, M., Yonan, A.L., Gilliam, T.C., Cantor, R.M. & Geschwind, D.H. (2005). Quantitative genome scan and ordered-subsets analysis of autism endophenotypes support language QTLs. *Mol Psychiatry*, 10, 747-757

Amaral, D.G., Schumann, C.M. & Nordahl, C.W. (2008). Neuroanatomy of autism. *Trends Neurosci*, 31, 137-145

Anney, R., Klei, L., Pinto, D., Regan, R., Conroy, J., Magalhaes, T.R. et al. (2010). A genome-wide scan for common alleles affecting risk for autism. *Hum Mol Genet*, 19, 4072-4082

Ashley-Koch, A.E., Mei, H., Jaworski, J., Ma, D.Q., Ritchie, M.D., Menold, M.M. et al. (2006). An analysis paradigm for investigating multi-locus effects in complex disease: examination of three $GABA_A$ receptor subunit genes on 15q11-q13 as risk factors for autistic disorder. *Ann Hum Genet*, 70, 281-292

Autism Genome Project Consortium, Szatmari, P., Paterson, A.D., Zwaigenbaum, L., Roberts, W., Brian, J., Liu, X.Q. et al. (2007). Mapping autism risk loci using genetic linkage and chromosomal rearrangements. *Nat Genet*, 39, 319-328

Azevedo, L., Suriano, G., van Asch, B., Harding, R.M. & Amorim, A. (2006). Epistatic interactions: how strong in disease and evolution? *Trends Genet*, 22, 581-585

Azevedo, R.B.R., Lohaus, R., Srinivasan, S., Dang, K.K. & Burch, C.L. (2006). Sexual reproduction selects for robustness and negative epistasis in artificial gene networks. *Nature*, 440, 87-90

Badcock, C. & Crespi, B. (2006). Imbalanced genomic imprinting in brain development: an evolutionary basis for the aethiology of autism. *J Evol Biol*, 19, 1007-1032

Baird, G., Simonoff, E., Pickles, A., Chandler, S., Loucas, T., Meldrum, D. et al. (2006). Prevalence of disorders of the autism spectrum in a population cohort of children in South Thames: the special needs and autism project (SNAP). *Lancet*, 368, 210-215

Baron, I.S. (2008). Autism spectrum disorder: complex, controversial, and confounding. *Neuropsychol Rev*, 18, 271-272

Baron-Cohen, S. (2002). The extreme male brain theory of autism. *Trends Cogn Sci*, 6, 248-254

Baron-Cohen, S. (2004). The cognitive neuroscience of autism. *J Neurol Neurosurg Psychiatry*, 75, 945-948

Baron-Cohen, S. (2006). The hyper-systemizing, assortative mating theory of autism. *Progr Neuropsychopharmacol Biol Psychiatry*, 30, 865-872

Beaudet, A.L. (2007). Autism: highly heritable but not inherited. *Nat Med*, 13, 534-536

Benfey, P.N. & Mitchell-Olds, T. (2008). From genotype to phenotype: systems biology meets natural variation. *Science*, 320, 495-497

Betancur, C. (2011). Etiological heterogeneity in autism spectrum disorders: more than 100 genetic and genomic disorders and still counting. *Brain Res*, 1380, 42-77

Boddaert, N., Barthélémy, C., Poline, J.B., Samson, Y., Brunelle, F. & Zilbovicius, M. (2005). Autism: functional brain mapping of exceptional calender capacity. *Br J Psychiatry*, 187, 83-86

Bradford, Y., Haines, J., Hutcheson, H., Gardiner, M., Braun, T., Sheffield, V. et al. (2001). Incorporating language phenotypes strengthens evidence of linkage to autism. *Am J Med Genet*, 105, 539-547

Brosius, J. & Kreitman, M. (2000). Eugenics: evolutionary nonsense? *Nat Genet*, 25, 253

Cannell, J.J. (2010). On the aetiology of autism. *Acta Paediatrica*, 99, 1128-1130

Casanova, M.F., Buxhoeveden, D.P., Switala, A.E. & Roy, E. (2002). Neuronal density and architecture (gray level index) in the brains of autistic patients. *J Child Neurol*, 17, 515-521

Casanova, M.F., van Kooten, I.A.J., Seitala, A.E., van Engeland, H., Heinsen, H., Steinbusch, H.W., et al. (2006). Minicolumnar abnormalities in autism. *Acta Neuropathol*, 112, 287-303

Casanova, M.F., Awitala, A.E., Trippe, J. & Fitzgerald, M. (2007). Comparative minicolumnar morphometry of three distinguished scientists. *Autism*, 11, 557-569

CDC. (2009). Autism and Developmental Disabilities Monitoring Network Surveillance Year 2006 Principal Investigators. Prevalence of autism spectrum disorders: Autism and Developmental Disabilities Monitoring Network, United States, 2006. *MMWR Surveill Summ*, 58, 1-20

Chen, G.K., Kono, N., Geschwind, D.H. & Cantor, R.M. (2006). Quantitative trait locus analysis of nonverval communication in autism spectrum disorder. *Mol Psychiatry*, 11, 214-220

Constantino, J.N. & Todd, R.D. (2000). Genetic structure of reciprocal social behavior. *Am J Psychiatry*, 157, 2043-2045

Constantino, J.N. & Todd, R.D. (2003). Autistic traits in the general population: a twin study. *Arch Gen Psychiatry*, 60, 524-530

Constantino, J.N. & Todd, R.D. (2005). Intergenerational transmission of subthreshold autistic traits in the general population. *Biol Psychiatry*, 57, 655-660

Coutinho, A.M., Sousa, I., Martins, M., Correia, C., Morgadinho, T., Bento, C. et al. (2007). Evidence for epistasis between SLC6A4 and ITGB3 in autism etiology and in the determination of platelet serotonin levels. *Hum Genet*, 121, 243-256

Craddock, N. & Owen, M.J. (2010). The Kraepelinian dichotomy - going, going...but still not gone. *Br J Psychiatry*, 196, 92-95

Dawson, M., Soulières, I., Gernsbacher, M.A. & Mottron, L. (2007). The level and nature of autistic intelligence. *Psychol Sci*, 18, 657-662

de Geus, E.J., Wright, M.J., Martin, N.G. & Boomsma, D.I. (2001). Genetics of brain function and cognition. *Behav Genet*, 6, 489-495

Depienne, C., Moreno-De-Luca, D., Heron, D., Bouteiller, D., Gennetier, A., Delorme, R. et al. (2009). Screening for genomic rearrangements and methylation abnormalities of the 15q11-q13 region in autism spectrum disorders. *Biol Psychiatry*, 66, 349-359

Durand, C.M., Betancur, C., Boeckers, T.M., Bockmann, J., Chaste, P., Fauchereau, F. et al. (2007). Mutations in the gene encoding the synaptic scaffolding protein SHANK3 are associated with autism spectrum disorders. *Nat Genet*, 39, 25-27

Duvall, J.A., Lu, A., Cantor, R.M., Todd, R.D., Constantino, J.N. & Geschwind, D.H. (2007). A quantitative trait locus analysis of social responsiveness in multiplex autism families. *Am J Psychiatry*, 164, 656-662

Dworzynski, K., Ronald, A., Hayiou-Thomas, M., Rijsdijk, F., Happé, F., Bolton, P.F. et al. (2007). Aetiological relationship between language performance and autistic-like traits in childhood: a twin study. *Int J Lang Commun Disord*, 42, 273-292

Feng, J., Schroer, R., Yan, J., Song, W., Yang, C., Bockholt, A. et al. (2006). High frequency of neurexin 1β signal peptide structural variants in patients with autism. *Neurosci Lett*, 409, 10-13

Fitzgerald, M. (2003). *Autism and creativity: is there a link between autism in men and exceptional ability?*, Brunner-Routledge, ISBN 978-1583912133, Hove and New York

Folstein, S.E. & Rosen-Sheidley, B. (2001). Genetics of autism: complex aetiology for a heterogeneous disorder. *Nat Rev Genet*, 2, 943-955

Fombonne, E. (2009). Epidemiology of pervasive developmental disorders. *Pediatr Res*, 65, 591-598

Frazier, T.W., Youngstrom, E.A., Sinclair, L., Kubu, C.S., Law, P., Rezai, A. et al. (2010). Autism spectrum disorders as a qualitatively distinct category from typical behavior in a large, clinically ascertained sample. *Assessment*, 17, 308-320

Freitag, C.M. (2007). The genetics of autistic disorders and its clinical relevance: a review of the literature. *Mol Psychiatry*, 12, 2-22

Frith, U. (2001). Mind blindness and the brain in autism. *Neuron*, 32, 969-979

Fullerton, J. (2006). New approaches to the genetic analysis of neuroticism and anxiety. *Behav Genet*, 36, 147-161

Gale, G.D., Yazdi, R.D., Khan, A.H., Lusis, A.J., Davis, R.C. & Smith, D.J. (2009). A genome-wide panel of congenic mice reveals widespread epistasis of behavior quantitative trait loci. *Mol Psychiatry*, 14, 631-645

Garber, K. (2007). Autism's cause may reside in abnormalities at the synapse. *Science*, 317, 190-191

Garcia-Villamisar, D. & Hughes, C. (2007). Supported employment improves cognitive performance in adults with autism. *J Intellect Disabil Res*, 51(Pt 2), 142-150

Gauthier, J., Spiegelman, D., Piton, A., Lafrenière, R.G., Laurent, S., St-Onge, J. et al. (2009). Novel *de novo* SHANK3 mutation in autistic patients. *Am J Med Genet B Neuropsychiatr Genet*, 150B, 421-424

Geschwind, D. (2007). Autism: searching for coherence. *Biol Psychiatry*, 62, 929-950

Gibson, G. (2009). Decanalization and the origin of complex disease. *Nat Rev Genet*, 10, 134-140

Gillberg, C., Billstedt, E., Sundh, V. & Gillberg, I.C. (2010). Mortality in autism: a prospective longitudinal community-based study. *J Autism Dev Disord*, 40, 352-357

Gjuvsland, A.B., Hayes, B.J., Omholt, S.W. & Carlborg, Ö. (2007). Statistical epistasis is a generic feature of gene regulatory networks. *Genetics*, 175, 411-420

Glessner, J.T., Wang, K., Cai, G., Korvatska, O., Kim, C.E., Wood, S. et al. (2009). Autism genome-wide copy number variation reveals ubiquitin and neuronal genes. *Nature*, 459, 569-573

Goldberg, A.D., Allis, C.D. & Bernstein, E. (2007). Epigenetics: a landscape takes shape. *Cell*, 128, 635-638

Happé, F. (1999). Autism: cognitive deficit or cognitive style? *Trends Cogn Sci*, 3, 216-222

Happé, F., Ronald, A. & Plomin, R. (2006). Time to give up on a single explanation for autism. *Nat Neurosci*, 9, 1218-1220

Harrison, R., Papp, B., Pál, C., Oliver, S.G. & Delneri, D. (2007). Plasticity of genetic interactions in metabolic networks of yeast. *Proc Natl Acad Sci U S A*, 104, 2307-2312

Hattori, J., Ogino, T., Abiru, K., Nakano, K., Oka, M. & Ohtsuka, Y. (2006). Are pervasive developmental disorders and attention-deficit/hyperactivity disorder distinct disorders? *Brain Dev*, 28, 371-374

Hill, W.G., Goddard, M.E. & Visscher, P.M. (2008). Data and theory point to mainly additive genetic variance for complex traits. *Plos Genetics*, 4, e1000008

Hobbs, K., Kennedy, A., DuBray, M., Bigler, E.D., Petersen, P.B., McMahon, W. et al. (2007). A retrospective fetal ultrasound study of brain size in autism. *Biol Psychiatry*, 62, 1048-1055

Hoekstra, R.A., Bartels, M., Verweij, C.J.H. & Boomsma, D.I. (2007). Heritability of autistic traits in the general population. *Arch Pediatr Adolesc Med*, 161, 372-377

Ijichi, S. & Ijichi, N. (2004). The prenatal autistic imprinting hypothesis: developmental maladaptation to the environmental changes between womb and the social world. *Med Hypotheses*, 62, 188-194

Ijichi, S. & Ijichi, N. (2007). Computerized lifelong mentoring support using robot for autistic individuals. *Med Hypotheses*, 68, 493-498

Ijichi, S., Ijichi, N., Ijichi, Y., Kawamura, Y., Hashiguchi, T. & Morioka, H. (2008). For others: epistasis and the evolutionary survival of an extreme tail of the quantitative distribution of autistic assets. *Med Hypotheses*, 70, 515-521

Ijichi, S., Ijichi, N., Ijichi, Y., Sameshima, H., Saeki, Y., Hall, W.W. & Morioka, H. (2011). Non-additive interactions between monomorphic and polymorphic loci: a theoretical explanation of the genetic underpinning for complex traits, In: *Advances in Medicine and Biology. Vol. 13.*, Berhardt, L.V., editor, 133-148, Nova Science (New York),
https://www.novapublishers.com/catalog/product_info.php?products_id=22873

Jiang, Y.H., Sahoo, T., Michaelis, R.C., Bercovich, D., Bressler, J., Kashork, C.D. et al. (2004). A mixed epigenetic/genetic model for oligogenic inheritance of autism with a limited role for UBE3A. *Am J Med Genet*, 131, 1-10

Jones, M.B. & Szatmari, P. (2002). A risk-factor model of epistatic interaction, focusing on autism. *Am J Med Genet*, 114, 558-565

Jorde, L.B., Hasstedt, S.J., Ritvo, E.R., Mason-Brothers, A., Freeman, B.J., Pingree, C. et al. (1991). Complex segregation analysis of autism. *Am J Hum Genet*, 49, 932-938

Kamp-Becker, I., Smidt, J., Ghahreman, M., Heinzel-Gutenbrunner, M., Becker, K. & Remschmidt, H. (2010). Categorical and dimensional structure of autism spectrum disorders: the nosologic validity of Asperger syndrome. *J Autism Dev Disord*, 40, 921-929

Keller, M.C. & Miller, G. (2006). Resolving the paradox of common, harmful, heritable mental disorders: which evolutionary genetic models work best? *Behav Brain Sci*, 29, 405-452

Lamb, J.A., Moore, J., Bailey, A. & Monaco, A.P. (2000). Autism: recent molecular genetic advances. *Hum Mol Genet*, 9, 861-868

Lauritsen, M.B. & Ewald, H. (2001). The genetics of autism. *Acta Psychiatr Scand*, 103, 411-27.

Leonard, H., Nassar, N., Bourke, J., Blair, E., Mulroy, S., de Klerk, N. et al. (2008). Relation between intrauterine growth and subsequent intellectual disability in a ten-year population cohort of children in Western Australia. *Am J Epidemiol*, 167, 103-111

Lepagnol-Bestel, A.M., Maussion, G., Boda, B., Cardona, A., Iwayama, Y., Delezoide,A.L. et al. (2008). SLC25A12 expression is associated with neurite outgrowth and is upregulated in the prefrontal cortex of autistic subjects. *Mol Psychiatry*, 13, 385-397

Lord, C., Cook, E.D., Leventhal, B.L. & Amaral, D.G. (2000). Autism spectrum disorders. *Neuron*, 28, 355-363

Maestrini, E., Paul, A., Monaco, A.P. & Bailey, A. (2000). Identifying autism susceptibility genes. *Neuron*, 28, 19-24

Malmberg, R.L. & Mauricio, R. (2005). QTL-based evidence for the role of epistasis in evolution. *Genet Res*, 86, 89-95

Manolio, T.A., Collins, F.S., Cox, N.J., Goldstein, D.B., Hindorff, L.A., Hunter, D.J. et al. (2009). Finding the missing heritability of complex diseases. *Nature*, 461, 747-753

Marcus, G.F. & Rabagliati, H. (2006). Genes and domain specificity. *Trends Cogn Sci*, 10, 397-398.

Markram, H., Rinaldi, E. & Markram, K. (2007). The intense world syndrome: an alternative hypothesis for autism. *Front Neurosci*, 1, 77-96

Marshall, C.R., Noor, A., Vincent, J.B., Lionel, A.C., Feuk, L., Skaug, J. et al. (2008). Structural variation of chromosomes in autism spectrum disorder. *Am J Hum Genet*, 82, 477-488

Maussion, G., Carayol, J., Lepagnol-Bestel, A.M., Tores, F., Loe-Mie, Y., Mibreta, U. et al. (2008). Convergent evidence identifying MAP/microtubule affinity-regulating kinase 1 (MARK1) as a susceptibility gene for autism. *Hum Mol Genet*, 17, 2541-2551

McCleery, J.P., Allman, E., Carver, L.J. & Dobkins, K.R. (2007). Abnormal magnocellular pathway visual processing in infants at risk for autism. *Biol Psychiatry*, 62, 1007-1014

McGovern, C. & Sigman, M. (2005). Continuity and change from early childhood to adolescence in autism. *J Child Psychol Psychiatry*, 46, 401-408

Mitchison, N.A. (2000). Polymorphism in regulatory gene sequences. *Genome Biol*, 2, comment2001.1-6

Moessner, R., Marshall, C.R., Sutcliffe, J.S., Skaug, J., Pinto, D., Vincent, J. et al. (2007) Contribution of SHANK3 mutations to autism spectrum disorders. *Am J Hum Genet*, 81, 1289-1297

Moldin, S.O., Rubenstein, J.L.R. & Hyman, S.E. (2006). Can autism speak to neuroscience? *J Neurosci*, 26, 6893-6896

Morris, K.V. (2005). siRNA-mediated transcriptional gene silencing: the potential mechanism and a possible role in the histone code. *Cell Mol Life Sci*, 62, 3057-3066

Nesse, R.M. & Williams, G.C. (1994). *Why we get sick: the new science of Darwinian medicine*, Vintage Books, ISBN 0-679-74674-9, New York

Palmert, M.R. & Hirschhorn, J.N. (2003). Genetic approaches to stature, pubertal timing, and other complex traits. *Mol Genet Metab*, 80, 1-10

Pauler, F.M., Koerner, M.V. & Barlow, D.P. (2007). Silencing by imprinted noncoding RNAs: is transcription the answer? *Trends Genet*, 23, 284-292

Persico, A.M. & Bourgeron, T. (2006). Searching for ways out of the autism maze: genetic, epigenetic and environmental clues. *Trends Neurosci*, 29, 349-358

Pinto, D., Pagnamenta, A.T., Klei, L., Anney, R., Merico, D., Regan, R., et al. (2010). Functional impact of global rare copy number variation in autism spectrum disorders. *Nature*, 466, 368-372

Plomin, R., Owen, M.J. & McGuffine, P. (1994). The genetic basis of complex human behaviors. *Science*, 264, 1733-1739

Plomin, R. & Kosslyn, S.M. (2001). Genes, brain and cognition. *Nat Neurosci*, 4, 1153-1155

Plomin, R., Haworth, C.M.A. & Davis, O.S.P. (2009). Common disorders are quantitative traits. *Nat Rev Genet*, 10, 872-878

Posserud, M.B., Lundervold, A.J. & Gillberg, C. (2006). Autistic features in a total population of 7-9-year-old children assessed by the ASSQ (Autism Spectrum Screening Questionnaire). *J Child Psychol Psychiatry*, 47, 167-175

Rapin I. (1997). Autism. (author reply). *N Engl J Med*, 337, 1556-1557

Rapin, I. & Katzman, R. (1998). Neurobiology of autism. *Ann Neurol*, 43, 7-14

Rapin, I. (1999). Autism in search of a home in the brain. *Neurology*, 52, 902-904

Reiersen, A.M., Constantino, J.N., Volk, H.E. & Todd, R.D. (2007). Autistic traits in a population-based ADHD twin sample. *J Child Psychol Psychiatry*, 48, 464-472

Rinaldi, T., Kulangara, K., Antoniello, K. & Markram, H. (2007). Elevated NMDA receptor levels and enhanced postsynaptic long-term potentiation induced by prenatal exposure to valproic acid. *Proc Natl Acad Sci U S A*, 104, 13501-13506

Ronald, A., Happé, F. & Plomin, R. (2005). The genetic relationship between individual differences in social and nonsocial behaviours characteristic of autism. *Dev Sci*, 8, 444-458

Ronald, A., Happé, F., Bolton, P., Butcher, L.M., Price, T.S., Wheelwright, S. et al. (2006a). Genetic heterogeneity between the three components of the autism spectrum: a twin study. *J Am Acad Child Adolesc Psychiatry*, 45, 691-699

Ronald, A., Happé, F., Price, T.S., Baron-Cohen, S. & Plomin, R. (2006b). Phenotypic and genetic overlap between autistic traits at the extremes of the general population. *J Am Acad Child Adolesc Psychiatry*, 45, 1206-1214

Ronald, A., Simonoff, E., Kuntsi, J., Asherson, P. & Plomin, R. (2008). Evidence for overlapping genetic influences on autistic and ADHD behaviours in a community twin sample. *J Child Psychol Psychiatry*, 49, 535-542

Rowe, H.C., Hansen, B.G., Halkier, B.A. & Kliebenstein, D.J. (2008). Biochemical netwerks and epistasis shape the Arabidopsis theliana metabolome. *Plant Cell*, 20, 1199-1216

Rzhetsky, A., Wajngurt, D., Park, N. & Zheng, T. (2007). Probing genetic overlap among complex human phenotypes. *Proc Natl Acad Sci U S A*, 104, 11694-11699

Santangelo, S.L. & Tsatsanis, K. (2005). What is known about autism: genes, brain, and behavior. *Am J Pharmacogenomics*, 5, 71-92

Schanen, N.C. (2006). Epigenetics of autism spectrum disorders. *Hum Mol Genet*, 15, R138-150

Scheuffgen, K., Happé, F., Anderson, M. & Frith, U. (2000). High "intelligence," low "IQ"? speed of processing and measured IQ in children with autism. *Dev Psychopathol*, 12, 83-90

Schmitz, C. & Rezaie, P. (2008). The neuropathology of autism: where do we stand? *Neuropahtol Appl Neurobiol*, 34, 4-11

Sebat, J. (2007). Major changes in our DNA lead to major changes in our thinking. *Nature Genet*, 39(Suppl), S3-5

Sebat, J., Lakshmi, B., Malhotra, D., Troge, J., Lese-Martin, C., Walsh, T. et al. (2007). Strong association of *de novo* copy number mutations with autism. *Science*, 316, 445-449

Simpson, D. (2003). Autism spectrum disorder is not as certain as implied. *BMJ*, 326, 986

Skuse, D.H. (2000). Imprinting, the X-chromosoem, and the male brain: explaining sex differences in the liability to autism. *Pediatr Res*, 47, 9-16

Skuse, D.H., Mandy, W.P.L. & Scourfield, J. (2005). Measuring autistic traits: heritability, reliability and validity of the Social and Communication Disorders Checklist. *Br J Psychiatry*, 187, 568-572

Skuse, D.H. (2007). Rethinking the nature of genetic vulnerability to autistic spectrum disorders. *Trends Genet*, 23, 387-395

Sung, Y.J., Dawson, G., Munson, J., Estes, A., Schellenberg, G.D. & Wijsman, E.M. (2005). Genetic investigation of quantitative traits related to autism: use of multivariate polygenic models with ascertainment adjustment. *Am J Hum Genet*, 76, 68-81

Sykes, N.H. & Lamb, J.A. (2007). Autism: the quest for the genes. *Expert Rev Mol Med*, 9, 1-15.

Tonge, B.J., Dissanayake, C. & Brereton, A.V. (1994). Autism: fifty years on from Kanner. *J Paediatr Child Health*, 30, 102-107

Tordjman, S., Ferrari, P., Sulmont, V., Duyme, M. & Roubertoux, P. (1997). Androgenic activity in autism. *Am J Psychiatry*, 154, 1626-1627

van Vliet, J., Oates, N.A. & Whitelaw, E. (2007). Epigenetic mechanisms in the context of complex diseases. *Cell Mol Life Sci*, 64, 1531-1538

Viding, E. & Blakemore, S.J. (2007). Endophenotype approach to developmental psychopathology: implications for autism research. *Behav Genet*, 37, 51-60

Weiss, L.A., Purcell, S., Waggoner, S., Lawrence, K., Spektor, D., Daly, M.J. et al. (2007). Identification of EFHC2 as a quantitative trait locus for fear recognition in Turner syndrome. *Hum Mol Genet*, 16, 107-113

Weiss, L.A., Shen, Y., Korn, J.M., Arking, D.E., Miller, D.T., Fossdal, R. et al. (2008). Association between microdeletion and microduplication at 16p11.2 and autism. *N Engl J Med*, 358, 667-675

Willis-Owen, S.A. & Flint, J. (2006). The genetic basis of emotional behaviour in mice. *European J Hum Genet*, 14, 721-728

Wing, L. (1997). The autistic spectrum. *Lancet*, 350, 1761-1766

Yan, J., Oliveira, G., Coutinho, A., Yang, C., Feng, J., Katz, C. et al. (2005). Analysis of the neuroligin 3 and 4 genes in autism and other neuropsychiatric patients. *Mol Psychiatry*, 10, 329-335

Yan, J., Noltner, K., Feng, J., Li, W., Schroer, R., Skinner, C. et al. (2008). Neurecin 1α structural variants associated with autism. *Neurosci Lett*, 438, 368-370

Zhang, X.S. & Hill, W.G. (2005). Genetic variability under mutation selection balance. *Trends Ecol Evol*, 20, 468-470

Zhao, X., Leotta, A., Kustanovich, V., Lajonchere, C., Geschwind, D.H., Law, K. et al. (2007). A unified genetic theory for sporadic and inherited autism. *Proc Natl Acad Sci U S A*, 104, 12831-12836

7

A New Genetic Mechanism for Autism

Julie Gauthier and Guy A. Rouleau
Université de Montréal
Canada

1. Introduction

It has been almost half a century since Leo Kanner first described the clinical phenotype associated with Autism[1]. Since Kanner's descriptions, much effort has been devoted to understanding and identifying the factors which may contribute to Autism. It was not until the early 80's that compelling evidence started to accumulate suggesting that Autism is a disorder of abnormal brain development. It is now generally accepted that both genetic and environmental factors are implicated in the etiology of this intriguing condition. In the current chapter, we will focus on the role of genetic factors involved in the pathogenesis of Autism. While multiple hypotheses have been proposed to account for the genetic origin of Autism, some have received more empirical support than others. While it is likely that more than one genetic mechanisms is involved in the pathogenesis of this disorder, this chapter will focus on a recent hypothesis implicating *de novo* mutations in synaptic genes. This hypothesis is based on the proposition that rare, highly penetrant mutations affecting any of many different genes which code for synaptic molecules, and which are specific to single families, predispose to Autism. Empirical lines of evidence for this hypothesis will be presented, along with examples, some of which are derived from work by our group.

1.1 The role of genetics in autism

Over the last two decades many studies aimed at identifying the genetic causes of Autism have shown that genetic factors play a predominant role in the genesis of this disorder. Twin studies predict that the heritability (i.e. the degree to which a given trait is controlled by inheritance) of Autism is between 70-90% (Bailey et al., 1995; Marco & Skuse, 2006; Lichtenstein et al., 2010). The relatively low concordance rate in dizygotic twins, the sharp decrease in recurrence risk of Autism in second- and third-degree relatives of autistic subjects (0.18% and 0.12% respectively) as well as a low risk in first-degree relatives of autistic subjects (3-7%) (Chakrabarti & Fombonne, 2001; Muhle et al., 2004) predicts two different genetic scenarios: 1) Autism could be explained by the co-inheritance in one individual of multiple disease-predisposing alleles, each with a small but additive effect, resulting in disease, or 2) by *de novo* mutations (i.e. not inherited from either parent). The first scenario is a polygenic inheritance, where the effect of multiple risk genes acting additively or multiplicatively results in disease. The second, very recently considered for Autism, argues that a fraction of the cases would result from incompletely penetrant new

---

[1] In this chapter the term Autism refers to the autism spectrum disorders which include Autism disorder, Asperger syndrome and Pervasive developmental disorder not otherwise specified

mutations or *de novo* mutations, such that only identical twins would share the genetic predisposition to Autism, hence the much higher monozygotic concordance than dizygotic concordance. Familial aggregation studies demonstrate that the risk of developing Autism is greater in offspring than in the parents. There are many possible explanations, one of which is that new mutations cause a fraction of Autism cases.

Dozens of genome-wide genetic linkage studies have been conducted in Autism kindred, identifying genetic signals of various significance levels on almost every arm of every human chromosome. There are few instances of consistent replication of linkage to any one site. Fine-mapping of these loci has been difficult, mainly because of genetic heterogeneity, so researchers have frequently opted to look at candidate genes directly based on the linkage results and/or on relevant gene functions. Association studies have also been used to search for genes in Autism. Although, in general, most studies have not been replicated, a few have been yielding a crop of possible susceptibility genes. Although the few replicated positive association studies are promising, it is surprising that no causative mutations or sequence variants have been identified in any of the loci associated with the disorder. In the absence of such mutations, the role of these genes in Autism remains unproven.

A number of genes have been strongly associated with Autism. For example, the X-linked neuroligin genes, *NLGN3* and *NLGN4*, two synaptic molecules, have been found to be associated with Autism. In the original report, two families with affected brothers (one with autism disorder and the other with Asperger syndrome) have a frameshifting and a missense mutation within the coding region of *NLGN4* and *NLGN3*, respectively (Jamain et al., 2003). Interestingly, both were new mutations that occurred in the mother of the affected brothers. Other mutations have since been described in these genes further supporting their role in the pathogenesis of Autism. *NLGN3* and *NLGN4* belong to the neuroligin family of postsynaptic cell adhesion molecules that are widely expressed in the brain (Philibert et al., 2000). The products of these genes are involved in late steps of synaptogenesis, mediating the specific recruitment of pre- and postsynaptic proteins to the site of initial synaptic contact. Two independent studies have shown *in vitro* that the frameshifting and the missense mutations described in humans alter the formation of presynaptic terminals (Chih et al., 2004; Comoletti et al., 2004). Since this discovery, mutations in other genes have been linked to Autism (discussed below). These findings support our hypothesis, which will be discussed in section below, that Autism is mainly a synaptic disorder largely caused by *de novo* mutations in synaptic genes.

2. The concept of "*de novo*" mutations

Recent studies on the direct measurement of human mutation rate have revealed that in any single conceptus there is approximately 1.1×10^{-8} (0.76×10^{-8} to 2.2×10^{-8}) mutation per base per generation (Awadalla et al., 2010; Lynch, 2010; Roach et al., 2010). A newborn is thought to have acquired about sixty new mutations in his/her genome. Among these, approximately 0.86 new deleterious mutation will lead to an altered amino acid, which corresponds to an average of about 1 new coding mutation per conceptus (Eyre-Walker & Keightley, 1999; Giannelli et al., 1999; Crow, 2000). *De novo* or spontaneous germline mutations can lead to serious clinical consequences, such as a disease, when affecting critical genes.

2.1 Common disease and common variants

Classical linkage and association studies, as mentioned earlier, have largely failed to identify predisposing genes for Autism as well as a number of other psychiatric disorders. The main reason for this lack of success is likely to be allelic and non-allelic genetic heterogeneity, with dozens to perhaps hundreds of genes predisposing to Autism, with each gene having many allelic variants. Such heterogeneity would require an enormous sample size to detect predisposing genes using population genetic approaches. It is likely that this heterogeneity results mostly from our limited ability to sub-phenotype brain disorders, particularly behavioural disabilities. The diagnosis of most psychiatric disorders remains largely based on clinical criteria, which define broad categories of dysfunction that may or may not be biologically linked. To date, there is no consistent, biologically validated method for defining these sub-phenotypes. Simply stated, Autism, as currently defined, probably result from so many different genes and alleles that classical genetic methods will prove inefficient in the identification of susceptibility genes for this disorder.

The hypothesis that a common disease may be caused by common variants was the favoured model for the genetic architecture of Autism until recently. Indeed, the constellation of published association studies reflects the widespread belief of the involvement of common variants in Autism. This hypothesis was appealing to many investigators since the common variants should be identifiable using methods such as linkage disequilibrium (Reich & Lander, 2001). Unfortunately, there are very few examples to support this hypothesis, particularly for brain disorders. Clinicians argue that Autism is a highly heterogeneous group of disorders, and none of them can be explained by single or even a few common variants. If this were the case, the plethora of genetic studies performed over the years should already have identified some of these variants. The widely distributed linkage and association positive signals scattered all over the genome rejects the existence of one or a few major predisposing common variants in this disorder. Furthermore, the few genes that have been found to definitely predispose to Autism explain only a small fraction of cases. This is not to say that some common variants will not be found for some predisposing genes, but this mechanism is unlikely to explain all the genetics of this condition.

It has recently been recognized that many complex disorders may result from a mix of common and rare variants. Let us consider breast cancer as an example (Nathanson et al., 2001): *BRCA1* and *BRCA2* genes contribute to a relatively common genetic disorder, but have many different rare mutations, even in a founder population (the Ashkenazim). For a complex trait such as Autism, the occurrence of many rare variants in many different disease predisposing genes seems to better predict the genetic architecture of the disorders (Pritchard, 2001; Pritchard & Cox, 2002; Smith & Lusis, 2002).

As a final point on the common disease common variant hypothesis, most studies looking at disease-causative mutations for Autism report mutations that are not recurrent, i.e. not observed more than once and specific to one individual. Again this suggests that mutations at many different loci may contribute to Autism, a result consistent with the failure to find common heritable variants with a major effect on disease risk. Lack of recurrence of mutations may in fact reflect the possibility that autistic traits can result from many different genetic defects.

2.2 Rare variants and new mutations

While not all amino acid substitutions will be deleterious, a significant fraction will be and may lead to disease. Therefore, for a disorder that may result from dysfunction in any one of hundreds of different genes, new mutations may be responsible for a significant fraction of cases. For example, should dysfunction in any of 100 different genes potentially lead to

Autism, and assuming amino acid changes lead to gene dysfunction in one fifth of instances, new mutations could cause Autism in one case out of 1,000 births, which would correspond to over 10% of cases based on the overall population incidence.
Looking at simple Mendelian traits, we can see that new mutations are common. For example, 1 in 6,000 live births harbour a novel mutation causing neurofibromatosis type 1 (Stephens et al., 1992; Grimm et al., 1994; Hudson et al., 1997). The frequency of new point mutations in Duchenne Muscular Dystrophy is similar, 1 in 10,500 live births (Grimm et al., 1994). One can argue that these are large genes, allowing for high mutation rate. Nonetheless, these are surprisingly high numbers of novel deleterious mutations. Let us consider Rett syndrome, which is closely related to Autism, and results from mutations in the relatively small *MECP2* gene (4 exons, 498 amino acids). The incidence of Rett syndrome is one in 10,000-15,000 females. Because 99-99.5% of all cases are sporadic with new mutations, this represents a new mutation rate of one in 5,000-7,500 live births for this small gene of 498 amino acids (Hagberg, 1985; Hagberg & Hagberg, 1997; Van den Veyver & Zoghbi, 2001). This example clearly shows that, for neurodevelopmental disorders, new mutations can act dominantly and can occur with a high enough frequency to explain the relatively high incidence of Autism. A high rate of new mutations can in part explain why genetic studies have so far failed to identify many Autism genes, and why diseases have been identified for a mere 3% of genes in the human genome. Mutations in genes leading to severe outcome where there is a strong negative selection against the phenotype, such as lethality in embryonic stages or reduced reproductive fitness, will not be transmitted to multiple family members, and therefore will not be detected by linkage gene mapping.

3. The role of *de novo* mutation in autism

Though we predict that *de novo* mutations will be a frequent cause of Autism, we do not think that it will be the only genetic explanation. The alternative genetic hypothesis for complex traits, mentioned previously, predicts that disease results from a combination or pattern of genotypes at different susceptibility loci. In recent years, statisticians have developed analytical methods that capture contributions from multiple susceptibility loci, and provide evidence for the localization of disease genes on human chromosomes (Sherriff & Ott, 2001; Hoh & Ott, 2003; Carlson et al., 2004). However, such analyses are very complex and yield few successful examples, even considering the simplest scenarios (Tiret et al., 1994; Bolk et al., 2000; Zetterberg et al., 2003). It seems that genome-wide searches using realistic sample sizes may not have the power to detect potential multi-gene interactions. The existence of new mutations, which contribute to this heterogeneity, makes classical genetic approaches even more difficult. Thus, the failure of conventional linkage and genome-wide association studies to identify but a few causative Autism genes is most likely due to two main confounding factors: phenotypic and genetic heterogeneity. Phenotypic heterogeneity is due to the inability to distinguish closely related clinical subtypes in the autism spectrum of behavioural disturbances. Genetic heterogeneity refers to fact that many different genes (and/or alleles of the same genes) lead to the same phenotype.

3.1 Monozygotic and dizygotic concordance

De novo mutations in identical twins would result in their sharing the same genetic predisposition to Autism. These alleles would be highly but not completely penetrant; hence the high monozygotic concordance and the low dizygotic concordance, as in the latter case the unaffected twin would not share the novel disease predisposing allele. Instances of non-penetrance would explain the fact that monozygotic twin concordance is not 100%. A *de novo* or spontaneous mutation can arise from different mechanisms and in different periods in the development of an individual. This kind of mutation can occur in the gametes (sperm or eggs), very early in the developing foetus or later in life as observed in cancer. The partial phenotypic concordance in monozygotic twin could also be in part explained by the occurrence of a *de novo* mutation early in the development of one twin, but not the other.

3.2 Reduced reproductive fitness

In the general population, the mutational load can be thought of as a balance between selection against a deleterious gene and its acquisition of new mutations. Lower rates of reproduction constitute a negative selection factor that should reduce the number of mutant alleles in the population, ultimately leading to decreased disease prevalence. These selective pressures tend to be of different intensity in different environments. In the case of Autism, only rarely do individuals with Autism have children, particularly the more severely affected individuals (Nicolson & Szatmari, 2003). Thus, Autism has a lower reproductive fitness (which is the ability to pass on genes by having offspring) due to an early age of onset and severely impaired cognitive and social functions. This observation should influence the disorders incidence and prevalence; but this is not what we observe. Autism incidence and prevalence seems to be relatively constant worldwide..

Studies of monogenic diseases[2] indicate that rare diseases with strong negative selection generally exhibit very large allelic diversity, hence many different mutations (Smith & Lusis, 2002). One exception to this pattern of high allelic diversity occurs when disease alleles also provide protection from negative environmental selective pressures. One example is the thalassemias, which confer resistance to malaria. Once arisen, the strong positive selective pressure conferred by these alleles allowed their relatively rapid spread through a specific population. However, such phenomena are usually regional, in response to specific regional environmental pressures. There are no examples of such phenomena occurring with equal strength in all cultural and geographical parts of the world, which needs to be the case to have a uniform incidence of Autism throughout the world. *De novo* mutations could explain this relatively uniform high incidence of disease, as new disease predisposing alleles will continually be introduced at a similar rate in all parts of the world.

3.3 Effects of paternal age

The male-to-female ratio of *de novo* mutations is estimated at about 4–6:1, presumably due to a higher number of germ-cell divisions with age in males(Crow, 2000). Therefore, one would predict that *de novo* mutations would more frequently come from males, particularly older men (Li et al., 2002). At the genetic level, increased risk for a disease with increasing paternal age can be explained by spermatogonial stem cell divisions that occur over the lifetime of males contributing to higher mutational rates in the sperm of older men. A higher paternal origin of *de novo* mutations has been shown for many diseases, including Apert syndrome (Moloney et al., 1996), Crouzon syndrome (Glaser et al., 2000), Multiple endocrine neoplasia type II (Carlson et al., 1994) and neurofibromatosis type 1 (Jadayel et al., 1990).

---

[2] Disorders caused by the inheritance of a single defective gene

Rett syndrome, a neurodevelopmental disease closely related to Autism, results almost entirely from new mutations which are exclusively of paternal origin (Girard et al., 2001; Trappe et al., 2001). A role for *de novo* mutations in Autism would predict that the incidence of disease should increase with increasing paternal age. Indeed, multiple recent studies reported advancing paternal age as a significant risk factor for Autism (Miller, 2006; Cantor et al., 2007; Croen et al., 2007; Puleo et al., 2008). Some authors have predicted a high incidence of male-derived novel mutations in many mental disorders (Preuss et al., 2004). Similar observation has been reported for schizophrenia (Malaspina et al., 2001) and intellectual disabilities (Malaspina et al., 2005), two conditions phenotypically related to Autism. These observations provide strong evidence that accumulation of *de novo* mutations in paternal sperm contributes to the overall risk of Autism.

3.4 Worldwide incidence

Data from a worldwide amalgam of studies show that the incidence of Autism has been maintained at a constant, relatively high prevalence in the worldwide population across a wide range of cultures and countries (McDonald & Paul, 2010). This occurs despite a strong negative selection against this condition. Indeed and with the exception of variants which date back to speciation, one would expect that common variants would result in a detectable uneven disease incidence across different populations due to migration, different population growth and isolation. This is not the case for Autism. In addition, this is not what one would predict in diseases with reduced reproductive fitness like Autism, unless there was a high new mutation rate. These observations emphasize the importance of *de novo* mutations in the pathogenicity of Autism.

Taken together, the high prevalence, the high monozygotic twin concordance, the predicted high level of allelic and non-allelic genetic heterogeneity, the uniform worldwide high incidence despite significantly reduced reproductive fitness, constitute evidences that Autism may result at least in part, from *de novo* mutations.

4. *De novo* mutations in genes associated with autism

The fact that a growing number of studies, several from our group, report the association of rare genetic variants with Autism constitutes strong evidence for the *de novo* hypothesis. Indeed, among causal genes identified for Autism, Rett syndrome and intellectual disability (three closely related disorders), the predisposing mutations, whether they be copy number variations, insertions/deletions or point mutations, are very frequently of *de novo* origin. A good example of such a gene is *SHANK3* encoding a synaptic scaffolding protein. Two of the first three mutations reported in the first manuscript linking *SHANK3* to Autism were actually of *de novo* origin; one a deletion of the terminal 22q13 and the other a G insertion leading to a frameshift that was carried by two affected brothers (Durand et al., 2007). None of these mutations were found in the parents. Many subsequent reports on mutation screening of *SHANK3* gene in Autim also find novel *de novo* mutations (Moessner et al., 2007; Gauthier et al., 2009). Other examples, such as the neuroligins genes, *NLGN3* and *NLGN4*, also clearly demonstrate the importance of *de novo* mutations in Autism. Jamain et al. found a single nucleotide insertion in two affected brothers, one with typical autism and the other with Asperger that arose *de novo* in the mother (Jamain et al., 2003). The *NRXN1* gene has also been found to harbour *de novo* pathogenic mutations in persons with Autism,

as well as in intellectual disabilities and in schizophrenia (Ching et al., 2010). Ching et al. found twelve deletions in *NRXN1* in patients with Autism and four were *de novo* copy number variations not identified in either parent (Ching et al., 2010). *SYNGAP1* (Hamdan et al.) and *IL1RAPL1* (Piton et al., 2008) are two other examples where we found *de novo* mutations in individuals with Autism and/or intellectual disability. Actually, *de novo* mutations are also a common cause of intellectual disability (Hamdan et al.; Vissers et al., 2010). As expected and as recently observed in Autism, *de novo* mutations have all been identified in different genes. For example, our group found six *de novo* deleterious mutations in females individuals with intellectual disability in *SYNGAP1* gene (encoding synaptic Ras GTPase activating protein 1) (Hamdan et al.; Piton et al., 2008; Hamdan et al., 2009).
In our recent study on the direct measurement of the *de novo* mutation rate in Autism and schizophrenia, we found a significant excess of potentially deleterious *de novo* mutations in individuals with Autism and schizophrenia (Awadalla et al., 2010). In this study, we examined variants identified by direct re-sequencing of 401 genes in a cohort of 285 autistic or schizophrenic individuals and for a subset of these genes in population control individuals. For the analysis, we distinguished functional from non-functional sites based on the effect of a mutation on the transcription or translation of the protein at a given position. Among trios[3] without family history of Autism, we observed a significant enrichment of functional *de novo* mutations ($p = 0.003$ in one-tail binomial test; $p = 0.022$ Fisher's exact test). Using a binomial test, our observed number of missense to nonsense *de novo* mutations was also significantly higher than the neutral expectation ($p = 0.04$), suggesting that some of the mutations are likely to be pathogenic. All of our reported observations suggest an excess of potentially disease-predisposing *de novo* mutations in the Autism and schizophrenia cohorts. Indeed, in this study, from sequencing only 8% of genes of the human genome, functional *de novo* mutations were found in 5% of individuals with no family history of Autism, exhibiting a wide range of clinical phenotypes. These few examples and many others recently published collectively provide strong evidence for a major role of *de novo* mutations in Autism

5. Altered synaptic connectivity in autism

The synapse is the locus of neural communication which is critical for human brain function. Defects in synaptic transmission are thought to underlie many common developmental brain disorders that are characterized by grossly normal brain structure (Zoghbi, 2003; Levitt et al., 2004). At a cellular level, there are presynaptic nerve endings specialized for the activity-dependent release of transmitter into the synaptic cleft, which is encapsulated by glial cells and contains adhesive molecules that keep presynaptic endings in register with postsynaptic specializations ("densities") on neural cell bodies and branches. In the mature nervous system these structures signal by chemical transmission and thus integrate and propagate the electrical signals that communicate through the brain. Synapses are thought to form in the embryo largely by genetically pre-programmed, activity-independent and evolutionarily conserved mechanisms (Goodman & Shatz, 1993). During post-natal development, which is the period during which many developmental brain diseases start to manifest themselves, synaptic activity is required to select, refine and stabilize mature connectivity patterns (Katz & Shatz, 1996). Thus cells that fire together wire together.

[3] A trio constitutes an affected individual and both his/her biological parents

Multiple indirect lines of evidence support the hypothesis of altered synaptic connectivity in Autism. These come in part from brain-imaging and neuropathological studies showing numerous alterations to both gross and microscopic structures of the brain of autistic individuals. For instance, an increased brain volume (Piven et al., 1996), increased brain weight (Bailey et al., 1998), abnormal neuronal morphology, with decreased complexity of dendritic branching and underdeveloped neuronal arbors (Bauman & Kemper, 1985; Raymond et al., 1996) have all been observed in autistic individuals. An abnormal neuronal density in the cerebellar hemispheres has also been observed (Bauman & Kemper, 1985). Notably, several components of the limbic system, including the amygdala (Lotspeich & Ciaranello, 1993) and the hippocampus (Raymond et al., 1996), have been found to be abnormal at the microscopic level. Cytoarchitectural features that are frequently abnormal include reduced numbers of Purkinje neurons in the cerebellum and vermis and small tightly packed neurons in regions of the limbic system, especially in the entorhinal cortex and in the medially placed nuclei of the amygdala. The reduced neuronal size and shortened dendritic pattern found in post-mortem studies are consistent with synaptic alterations. This synaptic deficiency hypothesis has been also proposed for schizophrenia, a neurodevelopmental disorder that is also characterized by marked disruptions of information processing and cognition (Glantz & Lewis, 2000). More recently, in an effort to directly determine if spine densities, or the synaptic connectivity, are altered in autistic subjects, Hussler and Zhang examined the structural microcircuitry within the cerebral cortex i.e. dendritic spine densities on cortical pyramidal cells from autistic subjects and age-matched control cases, on neurons located within both the superficial and deep cortical layers of frontal (BA 9), temporal (BA 21), and parietal lobe (BA 7). They observed several alterations in spine density in autistic subjects; for example the average spine densities in Autism were higher than those found in control cases, supporting altered synaptic connectivity and plasticity in the brains of individuals affected with Autism (Hutsler & Zhang, 2009).

Other evidence suggesting impaired synaptic function in autistic individuals includes the discovery of mutations in different synaptic genes, such as the neuroligins, the neurexins and *SHANK3* (see examples below in section 5.1). As mentioned earlier, Rett syndrome shares many features with Autism. Mutations in the coding region of the *MECP2* gene (a transcription repressor factor expressed by neurons and preferentially abundant in mature neurons) are known to be implicated in this severe disorder, with the vast majority of cases resulting from new mutations. Mutations in *MECP2* gene have also been identified in 3 females who meet the full diagnostic criteria for Autism, underscoring the similarity of these diseases (Carney et al., 2003). While the target genes for MECP2 protein remain unknown, the small brain size and the reduced neuronal and dendrite sizes in Rett Syndrome patients suggest that *MECP2* may play a role in synaptic processes (Shahbazian et al., 2002; Balmer et al., 2003). Recent findings using the olfactory system as a model to study MECP2 expression during development suggest that it may be involved in the formation of synaptic contacts (Cohen et al., 2003). These data further support the possibility that Autism results mainly from synaptic dysfunction.

5.1 Synaptic genes as candidates for autism

At a molecular level, synapses are organized as macromolecular "machines" (Grant, 2003). These synaptic machines consist of a presynaptic release apparatus and a signalling device at the postsynaptic density held together in quasi-crystalline registry at the adhesive cleft.

Many of the proteins constituting these various components have been identified by decades of synaptic biochemistry and physiological genetics, and their macromolecular assemblies have been characterized by proteomic analysis. The presynaptic release apparatus consists of proteins that include those for the structural cytoskeleton, vesicular membrane and trafficking components, vesicle fusion grid and nerve terminal membrane (Phillips et al., 2001; Blondeau et al., 2004). The postsynaptic density consists of structural proteins as well as signalling components such as tyrosine kinases and phosphatases, while both pre- and post-synaptic membranes contain fast voltage-gated channels and neurotransmitter-gated receptors, channels, transporters and G-protein coupled receptors mediating neuromodulation (Walikonis et al., 2000; Satoh et al., 2002) Rapid and selective communication across the synapse is ensured by the firm adhesion of each compartment at the cleft by cell surface as well as secreted extracellular matrix components (Huber et al., 2003). It is therefore not surprising that synaptic genes constitute the largest class of genes associated to developmental brain disorders - with many more to be discovered. Likewise, since many of these proteins are exposed at the extracellular surface, they could provide excellent "druggable" targets.

The discovery of genes clinically relevant to Autism is accelerating, with many involved in the synapse including several neuroligands, as well as genes involved in the glutamatergic pathway (Betancur et al., 2009). Of particular interest is the example of the synaptic cell adhesions and associated molecules including the neuroligins-neurexins-*SHANK3* genes. Mutations in the X-linked neuroligin-3 (*NLGN3*) and neuroligin-4 (*NLGN4X* and *NLGN4Y*) genes have been identified in brothers with autism. Laumonnier et al. identified a two base-pair deletion in *NLGN4* in 12 affected members of a French family with X-linked mental retardation, some of whom were also autistic (Laumonnier et al., 2004). Jamain et al. identified a C-to-T transition in the *NLGN3* gene, in two brothers, one with autism and the other with Asperger syndrome (Jamain et al., 2003). The *SHANK3* gene, which codes for a synaptic protein that binds directly to neuroligins, seems crucial for the development of language and social cognition. *SHANK3* mutations and small cytogenetic rearrangements have been implicated with the Autism phenotype (Durand et al., 2007; Gauthier et al., 2009). Other genes involved in this pathway have been found to be mutated in autistic individuals. Indeed, variants in *SHANK2 and LRRTM1* are reported in schizophrenia and Autism (Francks et al., 2007; Berkel et al., 2010). Other synaptic molecules implicated in Autism are the protocadherin family genes, which have been shown to be associated with Autism. Marshall et al. detected causative copy number variations in *PCDH9* gene and a *de novo* translocation deleting the *CDH18* genes in Autism (Marshall et al., 2008). Moreover, in a study of consanguineous families of Autism, a large homozygous deletion implicating *PCDH10* was detected (Morrow et al., 2008). All of these examples emphasized the role of impaired synaptic pathways in the pathogenesis of Autism.

6. Similar genetic architecture in other neurodevelopmental disorders

Autism, schizophrenia, and intellectual disability are all severe neurodevelopmental disorders that have childhood or early adulthood onset with a lifetime disability. Clinical manifestations of these disorders are diverse and complex, and include abnormalities in neuronal excitability, processing of complex information, as well as behaviors such as anxiety and impaired social interactions. Pathological studies, neuroimaging and other clinical observations predict that these disorders result from disrupted neurodevelopment caused by genetic and environmental factors (Lewis & Levitt, 2002). There is a significant overlap in clinical manifestations in these mental disorders, such as episodic psychosis and/or seizures, impaired cognitive functions, and language problems. Fifteen to thirty percent of Autism patients present with seizures and 20% of psychotic patients were diagnosed as having pervasive developmental disorders (Matese et al., 1994). Also, there is no clear clinical or neurobiological distinction between childhood schizophrenia, pervasive developmental disorder and autism (Mouridsen et al., 2000). Furthermore, these neurodevelopmental disorders can be included within the allelic spectrum of the same candidate gene. These observations strongly suggest that Autism, schizophrenia and intellectual disability may share similar pathogenic pathways and, thus, potential candidate genes. In addition, Autism, schizophrenia and intellectual disability have also a high prevalence, a high monozygotic twin concordance, a predicted high level of allelic and non-allelic genetic heterogeneity and a uniform worldwide high incidence despite significantly reduced reproductive fitness. All of these observations support the notion that the *de novo* mutations model be involved in all these disorders.

6.1 One gene, three phenotypic conditions

We believe that Autism, schizophrenia and intellectual disability, all severe neurodevelopmental disorders, can be studied in a similar manner, which focuses on the synaptic gene *de novo* mutation model. In addition, in some instances mutations in the same gene can lead to Autism, intellectual disability or schizophrenia, three clinically distinct phenotypes, as defined in the Diagnostic and Statistical Manual of Mental Disorders, the reference manual for the classification of mental disorders. A recent example is our findings with the *SHANK3* gene. Disruption of this synaptic gene was originally associated with the 22q13.3 deletion syndrome [OMIM 606232] characterized by neonatal hypotonia, global developmental delay, normal to accelerated growth, absent to severely delayed speech, autistic behavior (OMIM 209850), and minor dysmorphic features. In 2007, Durand et al. and Moessner et al., showed that abnormal gene dosage of SHANK3 is associated with Autism (Durand et al., 2007; Moessner et al., 2007). In addition, we identified a *de novo* splicing mutation in *SHANK3* in a patient with non-syndromic intellectual disability without Autism (Hamdan et al.). We also found deleterious *de novo* mutations in the *SHANK3* gene in a patient diagnosed with schizophrenia plus cognitive impairment. Similarly, mutations in the *NRXN*1 gene can lead to rare forms of Autism and schizophrenia (Kim et al., 2008; Rujescu et al., 2009). This phenomenon has been observed in other diseases and, as stressed by Zoghbi and Warren in their recent paper (Zoghbi & Warren, 2010), other examples include the *ARX* gene which causes X-linked lissencephaly, agenesis of corpus callosum with abnormal genitalia, cognitive deficits with or without seizures, or cognitive deficits, dystonia, and seizures; *LMNA* gene, which can lead to a diversity of disorders including Emery-Dreifuss muscular dystrophy Type 2, Charcot-Marie-Tooth axonal neuropathy limb girdle muscular dystrophy Type 1B, Hutchinson-Gilford progeria syndrome, and many other different clinical manifestations.

Altogether, these findings suggest that the neuroanatomical and physiological disturbances resulting from dysfunction of mutant genes may be influenced by the effect of genetic modifiers, the nature of the gene's role in the human brain and the effect of environmental experiences of the affected individuals, leading to different clinical outcomes in different patients. Differences in the mutation types (for example, point mutation vs. large gene disruptions) must certainly also contribute to the phenotypic variability. Although this observation is intriguing, multiple phenotypic manifestations from mutations of the same single gene have been described for many other diseases. Finally, the observation that one gene can lead to many phenotypes raised the question of whether Autism, schizophrenia and intellectual disability are different entities or part of a same phenotypic continuum.

7. Gene hunting approaches and the impact of the development of new technologies

In the last few years, a new generation of technologies, referred to as the next-generation DNA sequencing technologies have been developed which allow screening of the entire genome (i.e. > 20,000 genes) of single individuals within a matter of days. This new technology has revolutionized genetic research and has allowed new approaches in the search for diseases-causative genes. Before the advent of the next-generation DNA sequencing technologies, gene screening for the identification of disease-causative mutation was done one gene at a time. Next-generation DNA sequencing enable the parallel sequencing of all the 20,000 genes, leading to faster identification of mutations. These technologies therefore constitute the ideal method of screening for rare causative variants in all genes simultaneously. In the context of Autism and of the above mentioned *de novo* mutation genetic mechanism, the major advantage of these technologies is to make possible the identification of very rare *de novo* mutations, by comparing the genetic variants in an affected subject to those in both of his/her parents (a family trio). Given sufficient coverage and quality in next-generation DNA sequencing datasets, identifying *de novo* mutations in trios is highly feasible.

Sequencing of entire genomes is still rather expensive, so many groups now focus on the sequencing of the entire "exome" (i.e., the various coding regions of the genome) of an individual. Focusing on the exome is a reasonable approach as the vast majority of disease-causing mutations identified to date disrupt the protein-coding regions of genes. Such mutations include nonsense, small insertion/deletions, frameshifts, splicing and missense mutations, whose consequences can be predicted *in silico* based on well-annotated reference datasets (e.g. the human consensus CDS (CCDS) subset of the NCBI RefSeq database includes 23,339 consistently annotated protein-coding transcripts). These coding regions constitute less than 3% of the entire genome. Oligonucleotide hybridization-based methods that permit the capture and amplification of virtually all human exons at once, i.e. the human "exome", are now commercially available (NimbleGen, Agilent, Illumina). Limiting the analysis to the exome significantly increases the number of samples that can be sequenced, and is more likely to identify causative genes. Since it is likely that highly penetrant alleles (i.e. protein-truncating or missense) in different autistic cases will result from mutations in dozens of different genes, it is preferable to concentrate on re-sequencing only the coding regions of the genome, or the "exome", using targeted microarray capture followed by next generation sequencing. In addition mutations in coding regions are most easily interpreted thus making the link to the disorder easier to establish. The availability of these technologies is accelerating all aspects of the gene hunting process e.g. increasing number of genes that can be now analysed in a shorter period of time and the number of subjects being studied.

7.1 Challenges for the next-generation approaches

As next-generation DNA sequencing technologies improve, and as it becomes possible to rapidly produce detailed lists of variants per individual genome, the challenge will be to discriminate the pathogenic variants from the benign ones and establish the link to the disorder. Three major challenges can be identified. The first and most technical one is the ability to handle very large datasets in the order of tens or hundreds of terabytes in size, and access to powerful computing platforms that can process these datasets, and adequate resources for storage, retrieval and archiving. The second is the capacity to develop robust yet comprehensive methods to identify variants from next-generation DNA sequencing datasets. Choosing the correct sequence coverage and quality filters will ensure a maximum of true variants to be identified, with as few false positives or false negatives as possible. Several programs are available that can align short sequence reads to a reference genomic sequence and call potential homozygous or heterozygous variants. However, the total number of variants identified, even using different parameters within the same program, can vary largely. Intuitive paradigms and empirically determined cut off points need to be implemented. Furthermore, the accurate annotation of genomic variants is critical to classifying different variants for their potential impact on transcription, splicing or translation. This step must be comprehensive so that potential protein-truncating variants are not missed given that alternate splicing can lead to different transcripts with different open reading frames within a single gene. Finally, developing an experimental design based on the known or anticipated genetic mechanisms underlying the disease or condition, and on high quality diagnostic procedures, requires that affected and unaffected individuals be carefully selected from family groups to help prioritize variants for further analysis, and to maximize the chances of finding causative genes. In our case, identifying *de novo* mutations by analysing trios can very quickly lead to the identification of causative mutations and risk genes. Successful mastery of these key strategic and technical competencies is necessary to identify potentially pathogenic variants.

7.2 Copy number variations and autism

The use of microarray approaches for the detection of copy number variations, which its continuously improving resolution, provides additional evidence for the occurrence of *de novo* genomic events in the pathogenesis of Autism. In the last decade, studies linking copy number variation, and Autism have revealed that *de novo* and inherited copy number variations, including deletions and duplications, translocations, and inversions of chromosomes, all may significantly contribute to the pathogenesis of Autism, usually as penetrant rare variants (Sebat et al., 2007; Walsh et al., 2008). For example Sebat et al. showed that *de novo* copy number variations are more common in autistic patients than in non-autistic individuals (Sebat et al., 2007). They found that 10% of their patients with sporadic Autism (i.e. individuals with no history of the disorder) harboured a *de novo* copy number variation, while the frequency was only 3 % in patients with an affected first-degree relative and 1% in controls.

Other similar examples include the study from Marshall et al. (Marshall et al., 2008), Christian et al. (Christian et al., 2008) and Szatmari et al. (Szatmari et al., 2007) and more recently the report of Bremer et al. (Bremer et al.), which are all consistent with the hypothesis that *de novo* or weakly recurrent copy number variations seem to be significant contributing factor in the pathogenesis of Autism. Interestingly, based on the results of their copy number variations analyses, Marshall et al, concluded that "structural *de novo* were found in sufficiently high frequency in Autism subjects suggesting that cytogenetic and microarray analyses be considered in routine clinical workup" (Marshall et al., 2008). Although this is not the focus of this chapter, the genes identified by copy number variation analysis also support the notion that there are shared biological pathways in Autism, intellectual disability and schizophrenia (Guilmatre et al., 2009)

8. Conclusion and directions for future studies

As outlined throughout this chapter, several lines of evidence support the role of *de novo* mutations in the pathogenesis of Autism. *De novo* mutations are a well-established genetic mechanism for the development of a number of disorders such as Rett Syndrome and certain types of cancers but have been poorly explored for common diseases like Autism. In general, the development of technologies often brings new challenges, but mostly allows research to accelerate. Indeed, this technological progress is already starting to provide data supporting the role of *de novo* mutations in Autism. This hypothesis is gaining acceptance in the scientific community, as reflected in the growing number of recent publications on this subject. Although the focus of this chapter is on the role of *de novo* mutations in Autism, we acknowledge the fact that many other genetic or non-genetic mechanisms certainly contribute to this disorder.

The accessibility of next-generation DNA sequencing methodologies have enabled researchers to analyse a large amount of DNA and has had an important impact on gene hunting strategies , which have shifted from a tendency to look at single genes, one at the time, to multiple genes simultaneously. One interesting consequence of next-generation DNA sequencing is that it recently permitted to directly estimate the rate of *de novo* germline base substitution mutations in humans (Awadalla et al., 2010; Durbin et al., 2010). Based on these data, the challenge will be to determine if the observed *de novo* mutation rate detected in a disease is greater than the baseline rate.

In the last few years, researchers have identified several genes contributing to Autism, and most encode for proteins that are part of the synaptic machinery. An important concern for future research, where there will be rapid identification of many potential Autism gene mutations, will be to determine if they have a functional relevance to the disorder. Indeed, this question needs to be judiciously examined for most of the variants discovered. This will require to study model organism systems as proposed in our current Synapse to disease project (S2D; http://www.synapse2disease.ca), a large-scale medical research project launched in 2006 aiming to identify genes involved in several neurological and psychiatric diseases caused by defects in the development and functioning of the brain and nervous system. This project's philosophy is that once the base changes are discovered and considered likely to be "pathogenic mutations", biological validation must be conducted *in vitro* and *in vivo* in different model organism (e.g. fly, worm, fish, etc.) to determine their functional effects. Biological validation is an essential step often missing from most genetic studies and has thus severely limited data interpretation in the context of disease pathology. This validation will consequently help understanding pathways in neurodevelopmental disorders and ultimately give insights for the development of targeted therapeutic strategies.

Another challenge for future research in the field is the issue of whether genomic variants beyond the coding regions of a gene contribute to the etiology of the disorder. As mentioned earlier, the majority of mutations identified in Autism are located within coding regions but it should not be forgotten that variants in the non-coding regions, particularly the regulatory gene region, can also lead to disease.

Finally, the accessibility of the next-generation DNA sequencing technologies is facilitating the gene hunting process for researchers, whereas its application for clinical diagnostic testing seems to be inevitable, particularly as the cost per base continues to decrease. Although, the clinical tests based on these technologies represent particular challenges and will need careful validation, the connection between research findings in the genetics of Autism or any other neurodevelopmental disorders and clinical applications is closer than ever. All these research and technological advancements are for the greatest benefit of families.

9. Acknowledgments

We wish to thank Anna Bonnel and Fadi F. Hamdan for their critical reading of this manuscript.

10. References

Awadalla, P., J. Gauthier, R. A. Myers, F. Casals, F. F. Hamdan, A. R. Griffing, M. Cote, E. Henrion, D. Spiegelman, J. Tarabeux, A. Piton, Y. Yang, A. Boyko, C. Bustamante, L. Xiong, J. L. Rapoport, A. M. Addington, J. L. DeLisi, M. O. Krebs, R. Joober, B. Millet, E. Fombonne, L. Mottron, M. Zilversmit, J. Keebler, H. Daoud, C. Marineau, M. H. Roy-Gagnon, M. P. Dube, A. Eyre-Walker, P. Drapeau, E. A. Stone, R. G. Lafreniere&G. A. Rouleau (2010). "Direct measure of the de novo mutation rate in autism and schizophrenia cohorts." Am J Hum Genet 87(3): 316-24.

Bailey, A., A. Le Couteur, I. Gottesman, P. Bolton, E. Simonoff, E. Yuzda&M. Rutter (1995). "Autism as a strongly genetic disorder: evidence from a British twin study." Psychol Med 25(1): 63-77.

Bailey, A., P. Luthert, A. Dean, B. Harding, I. Janota, M. Montgomery, M. Rutter&P. Lantos (1998). "A clinicopathological study of autism." Brain 121 (Pt 5): 889-905.

Balmer, D., J. Goldstine, Y. M. Rao&J. M. LaSalle (2003). "Elevated methyl-CpG-binding protein 2 expression is acquired during postnatal human brain development and is correlated with alternative polyadenylation." J Mol Med 81(1): 61-8.

Bauman, M.&T. L. Kemper (1985). "Histoanatomic observations of the brain in early infantile autism." Neurology 35(6): 866-74.

Berkel, S., C. R. Marshall, B. Weiss, J. Howe, R. Roeth, U. Moog, V. Endris, W. Roberts, P. Szatmari, D. Pinto, M. Bonin, A. Riess, H. Engels, R. Sprengel, S. W. Scherer&G. A. Rappold (2010). "Mutations in the SHANK2 synaptic scaffolding gene in autism spectrum disorder and mental retardation." Nat Genet 42(6): 489-91.

Betancur, C., T. Sakurai&J. D. Buxbaum (2009). "The emerging role of synaptic cell-adhesion pathways in the pathogenesis of autism spectrum disorders." Trends Neurosci 32(7): 402-12.

Blondeau, F., B. Ritter, P. D. Allaire, S. Wasiak, M. Girard, N. K. Hussain, A. Angers, V. Legendre-Guillemin, L. Roy, D. Boismenu, R. E. Kearney, A. W. Bell, J. J. Bergeron&P. S. McPherson (2004). "Tandem MS analysis of brain clathrin-coated vesicles reveals their critical involvement in synaptic vesicle recycling." Proc Natl Acad Sci U S A 101(11): 3833-8.

Bolk, S., A. Pelet, R. M. Hofstra, M. Angrist, R. Salomon, D. Croaker, C. H. Buys, S. Lyonnet&A. Chakravarti (2000). "A human model for multigenic inheritance: phenotypic expression in Hirschsprung disease requires both the RET gene and a new 9q31 locus." Proc Natl Acad Sci U S A 97(1): 268-73.

Bremer, A., M. Giacobini, M. Eriksson, P. Gustavsson, V. Nordin, E. Fernell, C. Gillberg, A. Nordgren, A. Uppstromer, B. M. Anderlid, M. Nordenskjold&J. Schoumans "Copy number variation characteristics in subpopulations of patients with autism spectrum disorders." Am J Med Genet B Neuropsychiatr Genet 156(2): 115-24.

Cantor, R. M., J. L. Yoon, J. Furr&C. M. Lajonchere (2007). "Paternal age and autism are associated in a family-based sample." Mol Psychiatry 12(5): 419-21.

Carlson, C. S., M. A. Eberle, L. Kruglyak&D. A. Nickerson (2004). "Mapping complex disease loci in whole-genome association studies." Nature 429(6990): 446-52.

Carlson, K. M., J. Bracamontes, C. E. Jackson, R. Clark, A. Lacroix, S. A. Wells, Jr.&P. J. Goodfellow (1994). "Parent-of-origin effects in multiple endocrine neoplasia type 2B." Am J Hum Genet 55(6): 1076-82.

Carney, R. M., C. M. Wolpert, S. A. Ravan, M. Shahbazian, A. Ashley-Koch, M. L. Cuccaro, J. M. Vance&M. A. Pericak-Vance (2003). "Identification of MeCP2 mutations in a series of females with autistic disorder." Pediatr Neurol 28(3): 205-11.

Chakrabarti, S.&E. Fombonne (2001). "Pervasive developmental disorders in preschool children." JAMA 285(24): 3093-9.

Chih, B., S. K. Afridi, L. Clark&P. Scheiffele (2004). "Disorder-associated mutations lead to functional inactivation of neuroligins." Hum Mol Genet 13(14): 1471-7.

Ching, M. S., Y. Shen, W. H. Tan, S. S. Jeste, E. M. Morrow, X. Chen, N. M. Mukaddes, S. Y. Yoo, E. Hanson, R. Hundley, C. Austin, R. E. Becker, G. T. Berry, K. Driscoll, E. C. Engle, S. Friedman, J. F. Gusella, F. M. Hisama, M. B. Irons, T. Lafiosca, E. LeClair, D. T. Miller, M. Neessen, J. D. Picker, L. Rappaport, C. M. Rooney, D. P. Sarco, J. M. Stoler, C. A. Walsh, R. R. Wolff, T. Zhang, R. H. Nasir&B. L. Wu (2010). "Deletions of NRXN1 (neurexin-1) predispose to a wide spectrum of developmental disorders." Am J Med Genet B Neuropsychiatr Genet 153B(4): 937-47.

Christian, S. L., C. W. Brune, J. Sudi, R. A. Kumar, S. Liu, S. Karamohamed, J. A. Badner, S. Matsui, J. Conroy, D. McQuaid, J. Gergel, E. Hatchwell, T. C. Gilliam, E. S. Gershon, N. J. Nowak, W. B. Dobyns&E. H. Cook, Jr. (2008). "Novel submicroscopic chromosomal abnormalities detected in autism spectrum disorder." Biol Psychiatry 63(12): 1111-7.

Cohen, D. R., V. Matarazzo, A. M. Palmer, Y. Tu, O. H. Jeon, J. Pevsner&G. V. Ronnett (2003). "Expression of MeCP2 in olfactory receptor neurons is developmentally regulated and occurs before synaptogenesis." Mol Cell Neurosci 22(4): 417-29.

Comoletti, D., A. De Jaco, L. L. Jennings, R. E. Flynn, G. Gaietta, I. Tsigelny, M. H. Ellisman&P. Taylor (2004). "The Arg451Cys-neuroligin-3 mutation associated with autism reveals a defect in protein processing." J Neurosci 24(20): 4889-93.

Croen, L. A., D. V. Najjar, B. Fireman&J. K. Grether (2007). "Maternal and paternal age and risk of autism spectrum disorders." Arch Pediatr Adolesc Med 161(4): 334-40.

Crow, J. F. (2000). "The origins, patterns and implications of human spontaneous mutation." Nat Rev Genet 1(1): 40-7.

Durand, C. M., C. Betancur, T. M. Boeckers, J. Bockmann, P. Chaste, F. Fauchereau, G. Nygren, M. Rastam, I. C. Gillberg, H. Anckarsater, E. Sponheim, H. Goubran-Botros, R. Delorme, N. Chabane, M. C. Mouren-Simeoni, P. de Mas, E. Bieth, B. Roge, D. Heron, L. Burglen, C. Gillberg, M. Leboyer&T. Bourgeron (2007). "Mutations in the gene encoding the synaptic scaffolding protein SHANK3 are associated with autism spectrum disorders." Nat Genet 39(1): 25-27.

Durbin, R. M., G. R. Abecasis, D. L. Altshuler, A. Auton, L. D. Brooks, R. A. Gibbs, M. E. Hurles&G. A. McVean (2010). "A map of human genome variation from population-scale sequencing." Nature 467(7319): 1061-73.

Eyre-Walker, A.&P. D. Keightley (1999). "High genomic deleterious mutation rates in hominids." Nature 397(6717): 344-7.

Francks, C., S. Maegawa, J. Lauren, B. S. Abrahams, A. Velayos-Baeza, S. E. Medland, S. Colella, M. Groszer, E. Z. McAuley, T. M. Caffrey, T. Timmusk, P. Pruunsild, I. Koppel, P. A. Lind, N. Matsumoto-Itaba, J. Nicod, L. Xiong, R. Joober, W. Enard, B. Krinsky, E. Nanba, A. J. Richardson, B. P. Riley, N. G. Martin, S. M. Strittmatter, H. J. Moller, D. Rujescu, D. St Clair, P. Muglia, J. L. Roos, S. E. Fisher, R. Wade-Martins, G. A. Rouleau, J. F. Stein, M. Karayiorgou, D. H. Geschwind, J. Ragoussis, K. S. Kendler, M. S. Airaksinen, M. Oshimura, L. E. DeLisi&A. P. Monaco (2007). "LRRTM1 on chromosome 2p12 is a maternally suppressed gene that is associated paternally with handedness and schizophrenia." Mol Psychiatry 12(12): 1129-39, 1057.

Gauthier, J., D. Spiegelman, A. Piton, R. G. Lafreniere, S. Laurent, J. St-Onge, L. Lapointe, F. F. Hamdan, P. Cossette, L. Mottron, E. Fombonne, R. Joober, C. Marineau, P. Drapeau&G. A. Rouleau (2009). "Novel de novo SHANK3 mutation in autistic patients." Am J Med Genet B Neuropsychiatr Genet 150B(3): 421-4.

Giannelli, F., T. Anagnostopoulos&P. M. Green (1999). "Mutation rates in humans. II. Sporadic mutation-specific rates and rate of detrimental human mutations inferred from hemophilia B." Am J Hum Genet 65(6): 1580-7.

Girard, M., P. Couvert, A. Carrie, M. Tardieu, J. Chelly, C. Beldjord&T. Bienvenu (2001). "Parental origin of de novo MECP2 mutations in Rett syndrome." Eur J Hum Genet 9(3): 231-236.

Glantz, L. A.&D. A. Lewis (2000). "Decreased dendritic spine density on prefrontal cortical pyramidal neurons in schizophrenia." Arch Gen Psychiatry 57(1): 65-73.

Glaser, R. L., W. Jiang, S. A. Boyadjiev, A. K. Tran, A. A. Zachary, L. Van Maldergem, D. Johnson, S. Walsh, M. Oldridge, S. A. Wall, A. O. Wilkie&E. W. Jabs (2000). "Paternal origin of FGFR2 mutations in sporadic cases of Crouzon syndrome and Pfeiffer syndrome." Am J Hum Genet 66(3): 768-777.

Goodman, C. S.&C. J. Shatz (1993). "Developmental mechanisms that generate precise patterns of neuronal connectivity." Cell 72 Suppl: 77-98.

Grant, S. G. (2003). "Synapse signalling complexes and networks: machines underlying cognition." Bioessays 25(12): 1229-35.

Grimm, T., G. Meng, S. Liechti-Gallati, T. Bettecken, C. R. Muller&B. Muller (1994). "On the origin of deletions and point mutations in Duchenne muscular dystrophy: most deletions arise in oogenesis and most point mutations result from events in spermatogenesis." J Med Genet 31(3): 183-186.

Guilmatre, A., C. Dubourg, A. L. Mosca, S. Legallic, A. Goldenberg, V. Drouin-Garraud, V. Layet, A. Rosier, S. Briault, F. Bonnet-Brilhault, F. Laumonnier, S. Odent, G. Le Vacon, G. Joly-Helas, V. David, C. Bendavid, J. M. Pinoit, C. Henry, C. Impallomeni, E. Germano, G. Tortorella, G. Di Rosa, C. Barthelemy, C. Andres, L. Faivre, T. Frebourg, P. Saugier Veber&D. Campion (2009). "Recurrent rearrangements in synaptic and neurodevelopmental genes and shared biologic pathways in schizophrenia, autism, and mental retardation." Arch Gen Psychiatry 66(9): 947-56.

Hagberg, B. (1985). "Rett syndrome: Swedish approach to analysis of prevalence and cause." Brain Dev 7(3): 276-280.

Hagberg, B.&G. Hagberg (1997). "Rett syndrome: epidemiology and geographical variability." Eur Child Adolesc Psychiatry 6 Suppl 1: 5-7.

Hamdan, F. F., H. Daoud, A. Piton, J. Gauthier, S. Dobrzeniecka, M. O. Krebs, R. Joober, J. C. Lacaille, A. Nadeau, J. M. Milunsky, Z. Wang, L. Carmant, L. Mottron, M. H. Beauchamp, G. A. Rouleau&J. L. Michaud "De Novo SYNGAP1 Mutations in Nonsyndromic Intellectual Disability and Autism." Biol Psychiatry.

Hamdan, F. F., J. Gauthier, Y. Araki, D. T. Lin, Y. Yoshizawa, K. Higashi, A. R. Park, D. Spiegelman, S. Dobrzeniecka, A. Piton, H. Tomitori, H. Daoud, C. Massicotte, E. Henrion, O. Diallo, M. Shekarabi, C. Marineau, M. Shevell, B. Maranda, G. Mitchell, A. Nadeau, G. D'Anjou, M. Vanasse, M. Srour, R. G. Lafreniere, P. Drapeau, J. C. Lacaille, E. Kim, J. R. Lee, K. Igarashi, R. L. Huganir, G. A. Rouleau&J. L. Michaud "Excess of de novo deleterious mutations in genes associated with glutamatergic systems in nonsyndromic intellectual disability." Am J Hum Genet 88(3): 306-16.

Hamdan, F. F., J. Gauthier, D. Spiegelman, A. Noreau, Y. Yang, S. Pellerin, S. Dobrzeniecka, M. Cote, E. Perreault-Linck, L. Carmant, G. D'Anjou, E. Fombonne, A. M. Addington, J. L. Rapoport, L. E. Delisi, M. O. Krebs, F. Mouaffak, R. Joober, L. Mottron, P. Drapeau, C. Marineau, R. G. Lafreniere, J. C. Lacaille, G. A. Rouleau&J. L. Michaud (2009). "Mutations in SYNGAP1 in autosomal nonsyndromic mental retardation." N Engl J Med 360(6): 599-605.

Hoh, J.&J. Ott (2003). "Mathematical multi-locus approaches to localizing complex human trait genes." Nat Rev Genet 4(9): 701-9.

Huber, A. B., A. L. Kolodkin, D. D. Ginty&J. F. Cloutier (2003). "Signaling at the growth cone: ligand-receptor complexes and the control of axon growth and guidance." Annu Rev Neurosci 26: 509-63.

Hudson, J., C. L. Wu, M. Tassabehji, E. M. Summers, S. Simon, M. Super, D. Donnai&N. Thakker (1997). "Novel and recurrent mutations in the neurofibromatosis type 1 (NF1) gene." Hum Mutat 9(4): 366-367.

Hutsler, J. J.&H. Zhang (2009). "Increased dendritic spine densities on cortical projection neurons in autism spectrum disorders." Brain Res 1309: 83-94.

Jadayel, D., P. Fain, M. Upadhyaya, M. A. Ponder, S. M. Huson, J. Carey, A. Fryer, C. G. Mathew, D. F. Barker&B. A. Ponder (1990). "Paternal origin of new mutations in von Recklinghausen neurofibromatosis." Nature 343(6258): 558-559.

Jamain, S., H. Quach, C. Betancur, M. Rastam, C. Colineaux, I. C. Gillberg, H. Soderstrom, B. Giros, M. Leboyer, C. Gillberg&T. Bourgeron (2003). "Mutations of the X-linked genes encoding neuroligins NLGN3 and NLGN4 are associated with autism." Nat Genet 34(1): 27-9.

Katz, L. C.&C. J. Shatz (1996). "Synaptic activity and the construction of cortical circuits." Science 274(5290): 1133-8.

Kim, H. G., S. Kishikawa, A. W. Higgins, I. S. Seong, D. J. Donovan, Y. Shen, E. Lally, L. A. Weiss, J. Najm, K. Kutsche, M. Descartes, L. Holt, S. Braddock, R. Troxell, L. Kaplan, F. Volkmar, A. Klin, K. Tsatsanis, D. J. Harris, I. Noens, D. L. Pauls, M. J. Daly, M. E. MacDonald, C. C. Morton, B. J. Quade&J. F. Gusella (2008). "Disruption of neurexin 1 associated with autism spectrum disorder." Am J Hum Genet 82(1): 199-207.

Laumonnier, F., F. Bonnet-Brilhault, M. Gomot, R. Blanc, A. David, M. P. Moizard, M. Raynaud, N. Ronce, E. Lemonnier, P. Calvas, B. Laudier, J. Chelly, J. P. Fryns, H. H. Ropers, B. C. Hamel, C. Andres, C. Barthelemy, C. Moraine&S. Briault (2004). "X-linked mental retardation and autism are associated with a mutation in the NLGN4 gene, a member of the neuroligin family." Am J Hum Genet 74(3): 552-7.

Levitt, P., K. L. Eagleson&E. M. Powell (2004). "Regulation of neocortical interneuron development and the implications for neurodevelopmental disorders." Trends Neurosci 27(7): 400-6.

Lewis, D. A.&P. Levitt (2002). "Schizophrenia as a disorder of neurodevelopment." Annu Rev Neurosci 25: 409-32.

Li, S., D. Harrison, S. Carbonetto, R. Fassler, N. Smyth, D. Edgar&P. D. Yurchenco (2002). "Matrix assembly, regulation, and survival functions of laminin and its receptors in embryonic stem cell differentiation." J Cell Biol 157(7): 1279-1290.

Lichtenstein, P., E. Carlstrom, M. Rastam, C. Gillberg&H. Anckarsater (2010). "The genetics of autism spectrum disorders and related neuropsychiatric disorders in childhood." Am J Psychiatry 167(11): 1357-63.

Lotspeich, L. J.&R. D. Ciaranello (1993). "The neurobiology and genetics of infantile autism." Int Rev Neurobiol 35: 87-129.

Lynch, M. (2010). "Rate, molecular spectrum, and consequences of human mutation." Proc Natl Acad Sci U S A 107(3): 961-8.

Malaspina, D., S. Harlap, S. Fennig, D. Heiman, D. Nahon, D. Feldman&E. S. Susser (2001). "Advancing paternal age and the risk of schizophrenia." Arch Gen Psychiatry 58(4): 361-7.

Malaspina, D., A. Reichenberg, M. Weiser, S. Fennig, M. Davidson, S. Harlap, R. Wolitzky, J. Rabinowitz, E. Susser&H. Y. Knobler (2005). "Paternal age and intelligence: implications for age-related genomic changes in male germ cells." Psychiatr Genet 15(2): 117-25.

Marco, E. J.&D. H. Skuse (2006). "Autism-lessons from the X chromosome." Soc Cogn Affect Neurosci 1(3): 183-93.

Marshall, C. R., A. Noor, J. B. Vincent, A. C. Lionel, L. Feuk, J. Skaug, M. Shago, R. Moessner, D. Pinto, Y. Ren, B. Thiruvahindrapduram, A. Fiebig, S. Schreiber, J. Friedman, C. E. Ketelaars, Y. J. Vos, C. Ficicioglu, S. Kirkpatrick, R. Nicolson, L. Sloman, A. Summers, C. A. Gibbons, A. Teebi, D. Chitayat, R. Weksberg, A. Thompson, C. Vardy, V. Crosbie, S. Luscombe, R. Baatjes, L. Zwaigenbaum, W.

Roberts, B. Fernandez, P. Szatmari&S. W. Scherer (2008). "Structural variation of chromosomes in autism spectrum disorder." Am J Hum Genet 82(2): 477-88.

Matese, M., J. L. Matson&J. Sevin (1994). "Comparison of psychotic and autistic children using behavioral observation." J Autism Dev Disord 24(1): 83-94.

McDonald, M. E.&J. F. Paul (2010). "Timing of increased autistic disorder cumulative incidence." Environ Sci Technol 44(6): 2112-8.

Miller, M. C. (2006). "Older father, autistic child." Harv Ment Health Lett 23(6): 8.

Moessner, R., C. R. Marshall, J. S. Sutcliffe, J. Skaug, D. Pinto, J. Vincent, L. Zwaigenbaum, B. Fernandez, W. Roberts, P. Szatmari&S. W. Scherer (2007). "Contribution of SHANK3 mutations to autism spectrum disorder." Am J Hum Genet 81(6): 1289-97.

Moloney, D. M., S. F. Slaney, M. Oldridge, S. A. Wall, P. Sahlin, G. Stenman&A. O. Wilkie (1996). "Exclusive paternal origin of new mutations in Apert syndrome." Nat Genet 13(1): 48-53.

Morrow, E. M., S. Y. Yoo, S. W. Flavell, T. K. Kim, Y. Lin, R. S. Hill, N. M. Mukaddes, S. Balkhy, G. Gascon, A. Hashmi, S. Al-Saad, J. Ware, R. M. Joseph, R. Greenblatt, D. Gleason, J. A. Ertelt, K. A. Apse, A. Bodell, J. N. Partlow, B. Barry, H. Yao, K. Markianos, R. J. Ferland, M. E. Greenberg&C. A. Walsh (2008). "Identifying autism loci and genes by tracing recent shared ancestry." Science 321(5886): 218-23.

Mouridsen, S. E., B. Rich&T. Isager (2000). "A comparative study of genetic and neurobiological findings in disintegrative psychosis and infantile autism." Psychiatry Clin Neurosci 54(4): 441-6.

Muhle, R., S. V. Trentacoste&I. Rapin (2004). "The genetics of autism." Pediatrics 113(5): e472-86.

Nathanson, K. L., R. Wooster&B. L. Weber (2001). "Breast cancer genetics: what we know and what we need." Nat Med 7(5): 552-6.

Nicolson, R.&P. Szatmari (2003). "Genetic and neurodevelopmental influences in autistic disorder." Can J Psychiatry 48(8): 526-37.

Philibert, R. A., S. L. Winfield, H. K. Sandhu, B. M. Martin&E. I. Ginns (2000). "The structure and expression of the human neuroligin-3 gene." Gene 246(1-2): 303-10.

Phillips, G. R., J. K. Huang, Y. Wang, H. Tanaka, L. Shapiro, W. Zhang, W. S. Shan, K. Arndt, M. Frank, R. E. Gordon, M. A. Gawinowicz, Y. Zhao&D. R. Colman (2001). "The presynaptic particle web: ultrastructure, composition, dissolution, and reconstitution." Neuron 32(1): 63-77.

Piton, A., J. L. Michaud, H. Peng, S. Aradhya, J. Gauthier, L. Mottron, N. Champagne, R. G. Lafreniere, F. F. Hamdan, R. Joober, E. Fombonne, C. Marineau, P. Cossette, M. P. Dube, P. Haghighi, P. Drapeau, P. A. Barker, S. Carbonetto&G. A. Rouleau (2008). "Mutations in the calcium-related gene IL1RAPL1 are associated with autism." Hum Mol Genet 17(24): 3965-74.

Piven, J., S. Arndt, J. Bailey&N. Andreasen (1996). "Regional brain enlargement in autism: a magnetic resonance imaging study." J Am Acad Child Adolesc Psychiatry 35(4): 530-6.

Preuss, T. M., M. Caceres, M. C. Oldham&D. H. Geschwind (2004). "Human brain evolution: insights from microarrays." Nat Rev Genet 5(11): 850-60.

Pritchard, J. K. (2001). "Are rare variants responsible for susceptibility to complex diseases?" Am J Hum Genet 69(1): 124-37.

Pritchard, J. K.&N. J. Cox (2002). "The allelic architecture of human disease genes: common disease-common variant...or not?" Hum Mol Genet 11(20): 2417-23.

Puleo, C. M., A. Reichenberg, C. J. Smith, L. A. Kryzak&J. M. Silverman (2008). "Do autism-related personality traits explain higher paternal age in autism?" Mol Psychiatry 13(3): 243-4.

Raymond, G. V., M. L. Bauman&T. L. Kemper (1996). "Hippocampus in autism: a Golgi analysis." Acta Neuropathol 91(1): 117-9.

Reich, D. E.&E. S. Lander (2001). "On the allelic spectrum of human disease." Trends Genet 17(9): 502-10.

Roach, J. C., G. Glusman, A. F. Smit, C. D. Huff, R. Hubley, P. T. Shannon, L. Rowen, K. P. Pant, N. Goodman, M. Bamshad, J. Shendure, R. Drmanac, L. B. Jorde, L. Hood&D. J. Galas (2010). "Analysis of genetic inheritance in a family quartet by whole-genome sequencing." Science 328(5978): 636-9.

Rujescu, D., A. Ingason, S. Cichon, O. P. Pietilainen, M. R. Barnes, T. Toulopoulou, M. Picchioni, E. Vassos, U. Ettinger, E. Bramon, R. Murray, M. Ruggeri, S. Tosato, C. Bonetto, S. Steinberg, E. Sigurdsson, T. Sigmundsson, H. Petursson, A. Gylfason, P. I. Olason, G. Hardarsson, G. A. Jonsdottir, O. Gustafsson, R. Fossdal, I. Giegling, H. J. Moller, A. M. Hartmann, P. Hoffmann, C. Crombie, G. Fraser, N. Walker, J. Lonnqvist, J. Suvisaari, A. Tuulio-Henriksson, S. Djurovic, I. Melle, O. A. Andreassen, T. Hansen, T. Werge, L. A. Kiemeney, B. Franke, J. Veltman, J. E. Buizer-Voskamp, C. Sabatti, R. A. Ophoff, M. Rietschel, M. M. Nothen, K. Stefansson, L. Peltonen, D. St Clair, H. Stefansson&D. A. Collier (2009). "Disruption of the neurexin 1 gene is associated with schizophrenia." Hum Mol Genet 18(5): 988-96.

Satoh, K., M. Takeuchi, Y. Oda, M. Deguchi-Tawarada, Y. Sakamoto, K. Matsubara, T. Nagasu&Y. Takai (2002). "Identification of activity-regulated proteins in the postsynaptic density fraction." Genes Cells 7(2): 187-97.

Sebat, J., B. Lakshmi, D. Malhotra, J. Troge, C. Lese-Martin, T. Walsh, B. Yamrom, S. Yoon, A. Krasnitz, J. Kendall, A. Leotta, D. Pai, R. Zhang, Y. H. Lee, J. Hicks, S. J. Spence, A. T. Lee, K. Puura, T. Lehtimaki, D. Ledbetter, P. K. Gregersen, J. Bregman, J. S. Sutcliffe, V. Jobanputra, W. Chung, D. Warburton, M. C. King, D. Skuse, D. H. Geschwind, T. C. Gilliam, K. Ye&M. Wigler (2007). "Strong association of de novo copy number mutations with autism." Science 316(5823): 445-9.

Shahbazian, M. D., B. Antalffy, D. L. Armstrong&H. Y. Zoghbi (2002). "Insight into Rett syndrome: MeCP2 levels display tissue- and cell-specific differences and correlate with neuronal maturation." Hum Mol Genet 11(2): 115-24.

Sherriff, A.&J. Ott (2001). "Applications of neural networks for gene finding." Adv Genet 42: 287-97.

Smith, D. J.&A. J. Lusis (2002). "The allelic structure of common disease." Hum Mol Genet 11(20): 2455-61.

Stephens, K., L. Kayes, V. M. Riccardi, M. Rising, V. P. Sybert&R. A. Pagon (1992). "Preferential mutation of the neurofibromatosis type 1 gene in paternally derived chromosomes." Hum Genet 88(3): 279-282.

Szatmari, P., A. D. Paterson, L. Zwaigenbaum, W. Roberts, J. Brian, X. Q. Liu, J. B. Vincent, J. L. Skaug, A. P. Thompson, L. Senman, L. Feuk, C. Qian, S. E. Bryson, M. B. Jones, C.

R. Marshall, S. W. Scherer, V. J. Vieland, C. Bartlett, L. V. Mangin, R. Goedken, A. Segre, M. A. Pericak-Vance, M. L. Cuccaro, J. R. Gilbert, H. H. Wright, R. K. Abramson, C. Betancur, T. Bourgeron, C. Gillberg, M. Leboyer, J. D. Buxbaum, K. L. Davis, E. Hollander, J. M. Silverman, J. Hallmayer, L. Lotspeich, J. S. Sutcliffe, J. L. Haines, S. E. Folstein, J. Piven, T. H. Wassink, V. Sheffield, D. H. Geschwind, M. Bucan, W. T. Brown, R. M. Cantor, J. N. Constantino, T. C. Gilliam, M. Herbert, C. Lajonchere, D. H. Ledbetter, C. Lese-Martin, J. Miller, S. Nelson, C. A. Samango-Sprouse, S. Spence, M. State, R. E. Tanzi, H. Coon, G. Dawson, B. Devlin, A. Estes, P. Flodman, L. Klei, W. M. McMahon, N. Minshew, J. Munson, E. Korvatska, P. M. Rodier, G. D. Schellenberg, M. Smith, M. A. Spence, C. Stodgell, P. G. Tepper, E. M. Wijsman, C. E. Yu, B. Roge, C. Mantoulan, K. Wittemeyer, A. Poustka, B. Felder, S. M. Klauck, C. Schuster, F. Poustka, S. Bolte, S. Feineis-Matthews, E. Herbrecht, G. Schmotzer, J. Tsiantis, K. Papanikolaou, E. Maestrini, E. Bacchelli, F. Blasi, S. Carone, C. Toma, H. Van Engeland, M. de Jonge, C. Kemner, F. Koop, M. Langemeijer, C. Hijmans, W. G. Staal, G. Baird, P. F. Bolton, M. L. Rutter, E. Weisblatt, J. Green, C. Aldred, J. A. Wilkinson, A. Pickles, A. Le Couteur, T. Berney, H. McConachie, A. J. Bailey, K. Francis, G. Honeyman, A. Hutchinson, J. R. Parr, S. Wallace, A. P. Monaco, G. Barnby, K. Kobayashi, J. A. Lamb, I. Sousa, N. Sykes, E. H. Cook, S. J. Guter, B. L. Leventhal, J. Salt, C. Lord, C. Corsello, V. Hus, D. E. Weeks, F. Volkmar, M. Tauber, E. Fombonne, A. Shih&K. J. Meyer (2007). "Mapping autism risk loci using genetic linkage and chromosomal rearrangements." Nat Genet 39(3): 319-28.

Tiret, L., A. Bonnardeaux, O. Poirier, S. Ricard, P. Marques-Vidal, A. Evans, D. Arveiler, G. Luc, F. Kee, P. Ducimetiere&et al. (1994). "Synergistic effects of angiotensin-converting enzyme and angiotensin-II type 1 receptor gene polymorphisms on risk of myocardial infarction." Lancet 344(8927): 910-3.

Trappe, R., F. Laccone, J. Cobilanschi, M. Meins, P. Huppke, F. Hanefeld&W. Engel (2001). "MECP2 mutations in sporadic cases of Rett syndrome are almost exclusively of paternal origin." Am J Hum Genet 68(5): 1093-1101.

Van den Veyver, I. B.&H. Y. Zoghbi (2001). "Mutations in the gene encoding methyl-CpG-binding protein 2 cause Rett syndrome." Brain Dev 23 Suppl 1: S147-151.

Vissers, L. E., J. de Ligt, C. Gilissen, I. Janssen, M. Steehouwer, P. de Vries, B. van Lier, P. Arts, N. Wieskamp, M. del Rosario, B. W. van Bon, A. Hoischen, B. B. de Vries, H. G. Brunner&J. A. Veltman (2010). "A de novo paradigm for mental retardation." Nat Genet 42(12): 1109-12.

Walikonis, R. S., O. N. Jensen, M. Mann, D. W. Provance, Jr., J. A. Mercer&M. B. Kennedy (2000). "Identification of proteins in the postsynaptic density fraction by mass spectrometry." J Neurosci 20(11): 4069-80.

Walsh, T., J. M. McClellan, S. E. McCarthy, A. M. Addington, S. B. Pierce, G. M. Cooper, A. S. Nord, M. Kusenda, D. Malhotra, A. Bhandari, S. M. Stray, C. F. Rippey, P. Roccanova, V. Makarov, B. Lakshmi, R. L. Findling, L. Sikich, T. Stromberg, B. Merriman, N. Gogtay, P. Butler, K. Eckstrand, L. Noory, P. Gochman, R. Long, Z. Chen, S. Davis, C. Baker, E. E. Eichler, P. S. Meltzer, S. F. Nelson, A. B. Singleton, M. K. Lee, J. L. Rapoport, M. C. King&J. Sebat (2008). "Rare Structural Variants Disrupt

Multiple Genes in Neurodevelopmental Pathways in Schizophrenia." Science 320(5875): 539-543.

Zetterberg, H., A. Zafiropoulos, D. A. Spandidos, L. Rymo&K. Blennow (2003). "Gene-gene interaction between fetal MTHFR 677C>T and transcobalamin 776C>G polymorphisms in human spontaneous abortion." Hum Reprod 18(9): 1948-50.

Zoghbi, H. Y. (2003). "Postnatal neurodevelopmental disorders: meeting at the synapse?" Science 302(5646): 826-30.

Zoghbi, H. Y.&S. T. Warren (2010). "Neurogenetics: advancing the "next-generation" of brain research." Neuron 68(2): 165-73.

8

Common Genetic Etiologies and Biological Pathways Shared Between Autism Spectrum Disorders and Intellectual Disabilities

Liana Kaufman[1], Abdul Noor[1], Muhammad Ayub[2,3] and John B. Vincent[1,4]
[1]*Neurogenetics Section, The Centre for Addiction & Mental Health, Toronto, Ontario*
[2]*Tees, Esk and Wear Valleys NHS Foundation Trust*
[3]*School for Medicine and Health, University of Durham*
Wolfson Research Institute Queen's Campus, Stockton on Tees
[4]*Department of Psychiatry, University of Toronto, Toronto, Ontario*
[1,4]*Canada*
[2,3]*United Kingdom*

1. Introduction

1.1 Definitions of ID and autism

Intellectual disability (ID) is a common neurodevelopmental disorder that is characterized by an intelligence quotient (IQ) lower than 70, and deficits in at least two behaviors related to adaptive functioning diagnosed by 18 years of age (American Psychiatric Association, 2000). Adaptive functioning behaviours can be defined as the ability to acquire skills that help an individual to live independently and to cope with everyday life, and involves skills such as language/communication, social skills, home living and safety. ID ranges in its severity and may either be present co-morbidly with many congenital syndromes, or may present alone.

Autism is a severe, lifelong neurodevelopmental disorder characterized by impairments in three major domains: communication, socialization, and repetitive behavior. Leo Kanner first described this developmental disorder in 1943 as a social disorder featuring the innate inability to form typical, affective contact with others (Kanner, 1943). It is now known that the Autism Spectrum Disorder (ASD) includes a widely variable range of clinical phenotypes that have been grouped into individual disorders. Autistic disorder (autism), Asperger syndrome, pervasive developmental disorder not otherwise specified (PDD-NOS), Rett syndrome and child disintegrative disorder (CDD) are all currently separate diagnoses in the DSM-IV under the umbrella title "Pervasive Developmental Disorders"(American Psychiatric Association, 2000). These disorders differ from each other with regard to severity of symptoms, early development of language, deterioration in skills once they have developed, and cognitive development.

Individuals with autism show severe deficits in all three major phenotypic domains and present with abnormal development before age 3 years. In addition, cognitive functioning is frequently delayed. Individuals with Broad Autism Phenotype (BAP) have some symptoms of autism, but do not meet the full criteria for autism or ASD (Hurley et al., 2007). BAP is currently not a recognized diagnosis within the DSM-IV, however, it can be useful when looking at familiality in ASDs.

Asperger syndrome is characterized by qualitative impairment in social interaction and restricted repetitive and stereotyped patterns of behavior, interests and activities. Language and cognitive development is relatively unaffected in these individuals, however pragmatic or social language is often delayed (McConachie & Diggle, 2007). Individuals with PDD-NOS meet autism criteria and have later age of onset. These individuals may also show severe and pervasive impairment in one or two of the three core domains with or without cognitive or language delay. Rett syndrome occurs almost exclusively occurs in females and it is characterized by developmental arrest between 6 and 18 months of age, followed by loss of speech, stereotypical movements, microcephaly, seizures, and intellectual disability (Hagberg et al., 1983). CDD is characterized by normal development for at least the first two years (but up to 10 years) of life, followed by clinically significant loss of acquired skills. Individuals with CDD must not meet the diagnostic criteria for Rett syndrome or schizophrenia. They also must display at least two of the three diagnostic domains for autism.

While ASDs are often associated with ID, it should be noted that not all individuals with ASD have cognitive deficits. The current DSM-IV statistic states that 75% of individuals with ASD have some degree of intellectual disability. However, this may be an outdated over-estimate. More recent studies suggest an increasingly modest level; approximately 50-60% [71%(Chakrabarti & Fombonne, 2001) 63%(Bertrand et al., 2001); 40%(Baird et al., 2006), 50%(Charman et al., 2011]. With increased awareness in the general population and a greater understanding of the problem by professionals, an increasing number of people with high functioning Autism and Asperger syndrome are diagnosed with the condition. This has probably changed the proportion of cases with identifiable ID among the ASD population. It is also possible that expanding the diagnostic criteria for ASDs, which historically were much more limited, might explain this decrease (Charman et al., 2011).

The new DSM-V, which will be available in 2013, will define ASDs differently than the current version. The three behavioural domains will be condensed into two: Social communication and repetitive behaviours/narrow interests. In addition, the separate diagnoses within the spectrum will be removed. Autism, Rett syndrome, Asperger syndrome, PDD-NOS and CDD will all be classified as "autism spectrum disorder" with a specified etiology/syndrome if known (e.g. ASD with Rett syndrome; DSM-5 Proposed Revisions, 2010). In addition, a level of severity will be assigned to each diagnosis, which will reflect the functional level of the individual. The alteration in nomenclature reflects the clinical and interventional needs of the individuals and is intended to be more reflective of what is known of the pathology of the disorder (DSM-5 Proposed Revisions, 2010).

1.2 Classification of ID by IQ and syndromic vs. nonsyndromic

ID is currently subdivided into 5 categories based on intelligence quotient (IQ): mild, moderate, severe, profound and unable to classify (American Psychiatric Association, 2000; Table 1). However, epidemiological studies often use a simplified classification, grouping their subjects into mild ID (IQ50-70) and severe ID (IQ<50; Ropers & Hamel, 2005). IQ tests are a set of tasks which are administered to representative population samples for creation of norms. An IQ two standard deviations below the mean or lower is indicative of ID. The distribution of IQ in the population is normal in the main, apart from an increased number of cases in the tail on the lower end of IQ. On that basis, the population prevalence of ID should be close to 3% at least. The studies looking at prevalence of ID were recently reviewed systematically. The range reported varied between 0.93 per 1000 and 156.03 per 1000 (Maulik et al., 2011). Differences in rates of mild ID mainly account for this variation. The reasons are differences in definition of ID criteria, characteristics and setting of the sample studied and differences in methodology. The prevalence of severe ID is relatively stable (3-4 per 1000; (Roeleveld et al., 1997; Leonard & Wen, 2002; Emerson, 2007).

Severity	IQ	Proportion of ID	Functional Level
Borderline	70-84	N/A	Can live independently; May require low level support
Mild	50-69	85%	Can often live independently with social support
Moderate	35-49	10%	Acquire some communication and self-help skills, require moderate supervision
Severe	20-34	3-4%	Acquire only basic self-help and communication skills, require supervision
Profound/ Unspecified	<20	1-2%	Require highly structured and supervised living conditions

Table 1. This table indicates the categories of intellectual disabilities by IQ and ability to function in society as indicated by the DSM-IV-TR.

The new DSM-5 will likely be changing the severity criteria to encompass functional behavioural deficits and the level of interference that these have in the lives of affected individuals (DSM-5 Proposed Revisions, 2010). Although the new criteria have not yet been established, it will likely echo the changes to the criteria for autism, focusing on functional level as opposed to strict IQ cut-offs. The manual will also be changing the wording of the ID diagnostic criteria to encourage cultural sensitivity and relevance, and to ensure that culturally validated psychometric tests are used to evaluate IQ and level of functioning.

In addition to categorization by severity/IQ level, ID can also be grouped into syndromic intellectual disability (S-ID) and non-syndromic intellectual disability (NS-ID). In S-ID, individuals present with an identifiable constellation of clinical features or co-morbidities in addition to ID. While S-ID has a clear definition, there is debate over the classification of NS-ID. Traditionally, NS-ID has been defined by the presence of intellectual disability as the sole clinical feature. However, it has been a challenge to rule out the presence of more subtle physical signs, neurological anomalies and psychiatric disorders in these individuals, as they may be less apparent, or difficult to diagnose due to the cognitive impairment. Additionally, the symptoms of some syndromes may be so subtle that they are extremely difficult to diagnose unless the features are looked for specifically in the context of a known genetic defect previously associated with these features (Ropers, 2006). Thus the distinction between S-ID and NS-ID is often blurred.

1.3 Endophenotypes and "essential autism" versus "complex autism"

The extreme phenotypic heterogeneity of ASD poses a challenge for the study of underlying etiologies. It has been argued that delineation of the clinical heterogeneity of ASD may help in the identification of more homogeneous sub-groups for the study of etiological factors, and to predict the outcome and treatment choices. While ASD has three core phenotypic domains, it can also be sub-grouped on the basis of presence or absence of certain clinical features, termed endophenotypes. Endophenotypes of ASDs include IQ level, seizures, brain malformations, dysmorphology and head circumference (Viding & Blakemore, 2007). Historically, based on non-verbal IQ testing, ~70% of autistic children were reported as having some form of ID (Fombonne, 2003). While it is now thought that this number may be lower, it is essential to look at IQ as an endophenotype of autism when predicting outcomes. Previously published longitudinal studies report that IQ scores can strongly predict long-term outcomes of ASD and are directly associated with the psychopathology of autism, even in young children (Howlin et al., 2004). In addition, preschool cognitive functioning has been found to be a strong predictor of school-age functioning, and high IQ has been shown to be necessary but not sufficient for optimal outcome in the presence of severe language impairment (Stevens et al., 2000).

Another common central nervous system (CNS) dysfunction associated with autism is the high risk of epilepsy (Spence & Schneider, 2009). The prevalence of seizures in autism is estimated to be up to 46% (Hughes & Melyn, 2005) and it has been estimated that as many as 32% of individuals with epilepsy may meet the diagnostic criteria for ASD (Clarke et al., 2005). Notably, the prevalence of seizures is higher among autistic individuals with moderate to severe ID and individuals with overt motor abnormalities (Tuchman & Rapin, 2002). Furthermore, individuals with autism plus epilepsy on average have lower IQs. Epilepsy is one of the negative factors contributing to cognitive, adaptive and behavioral/emotional outcomes for autistic individuals (Hara, 2007).

Structural brain malformations, including accentuated Virchow–Robin space, acrocallosal syndrome and polymicrogyria have been reported to be associated with autism (Steiner et al., 2004; Schifter et al., 1994; Zeegers et al., 2006), however, until recently, MRIs have been considered to be of indeterminate value and they are not included in the standard clinical evaluation of autism. A recent study has revealed an unexpectedly high prevalence of brain abnormalities (48%) in autism patients. Some common abnormalities include white-matter signal abnormalities, severely dilated Virchow-Robin space and temporal lobe structural abnormalities (Boddaert et al., 2009).

Generalized dysmorphology, an indicator of insult in early development, has been reported in 15% to 20% of individuals with autism (Miles & Hillman, 2000) and has been proposed to be a predictor of a poor response to early intensive behavioral intervention. According to the Autism Dysmorphology Measure (ADM) guidelines, the 12 body areas assessed for dysmorphology are: height, hair growth pattern, structure and size of ear, nose size and shape, face size and structure, philtrum, mouth and lips, teeth, hand size, fingers and thumbs, nails and feet. The ADM was developed by the Miles laboratory at the University of Missouri to aid clinicians who are not extensively trained in medical genetics to distinguish between individuals with ASD with and without dysmorphisms (Miles & Hillman, 2000). Besides generalized dysmorphology, head size abnormalities (microcephaly and macrocephaly) have also been found in autistic individuals. Microcephaly, head circumference $<2^{nd}$ centile, occurs in 5 to 15% of children with autism and is a predictor of poor outcome (Miles et al., 2005). On the other hand, macrocephaly, head circumference

>97th centile has been observed in ~30% of children with autism (Miles et al., 2000). Generalized dysmorphology and microcephaly have been proposed as good predictors of clinical outcome and may be used to classify the autism phenotype into subgroups: complex autism and essential autism. Complex autism consists of autistic individuals who show evidence of some abnormality in early morphogenesis, manifested by either significant dysmorphology or microcephaly. The remainder without dysmorphic features or microcephaly are classified as having essential autism (Miles et al., 2005), and make up 70-80% of the autism population.

1.4 Diagnostic approaches

The symptoms of most ASDs are usually present by the age of three and may persist throughout the lifespan; however, CDD and PDD-NOS may present later. A number of checklists and diagnostic tools are available for diagnosis of autism. The Childhood Autism Rating Scale (CARS; Schopler et al., 1980) is a commonly used diagnostic checklist which consists of 15 questions scored by the parents and the examiner. It is a reliable and efficient tool that is commonly used in clinics. Another autism screening tool widely used in clinical settings is the Checklist for Autism in Toddlers-modified (M-CHAT). This checklist consists of 23 yes/no questions and is a promising tool for the early diagnosis of autism (Robins et al., 2001). Other such checklists include the Autism Behaviour Checklist (ABC; Witwer & Lecavalier, 2007) and Gilliam Autism Rating Scale (GARS; South et al., 2002).

The Autism Diagnostic Observation Schedule (ADOS) and Autism Diagnostic Interview-Revised (ADI-R) are two widely accepted instruments used for diagnosis of ASD in both clinical and research settings. ADI-R, a revised version of Autism Diagnostic Interview (ADI), is a semi-structured, investigator-based interview for the caregivers of children with autism and adults for whom autism or ASD is a possible diagnosis (Lord et al., 1994). The ADOS is a semi-structured, standardized assessment of social interactions, communication, play, and imaginative use of objects for children suspected of having ASD (Lord et al., 2000). It is an observational assessment of the child's behaviour, often performed by a psychologist or another trained professional. Checklist tools are widely used in clinical practice because of their ease and efficiency; however, ADI-R and ADOS have been adapted, particularly in recent years, to make them more appropriate for use in clinical settings, as well as for diagnosis of toddlers and patients with intellectual disabilities. In particular, the shorter version of ADOS is becoming increasingly popular in clinics.

All of these diagnostic tools have their own advantages and disadvantages. For example, ADI-R and ADOS are lengthy, require elaborate training and are suitable for use in more specialized settings. The Checklist for Autism in Toddlers-modified (M-CHAT), described earlier in this section, has been particularly useful in the frontline clinical world, as well as for early diagnosis of autism (Robins et al., 2001), and has overcome some of the challenges faced by ADI-R and ADOS. M-CHAT includes a checklist of 23 items to be filled out by parents and it can be administered at a much earlier stage to identify toddlers who are at risk of autism. A recent study has confirmed the validity of this instrument in detecting possible ASD at 16-30 months of age (Kleinman et al., 2008). However, the high sensitivity of this checklist means that some children without autism will fail the screening. It has been suggested that children who fail and do not have autism are at increased risk for other developmental disorders or delays and should be monitored accordingly (Robins et al., 2001).

Over last two decades the diagnosis of syndromes based on behavioural symptoms has become relatively standardized, although there are some issues around the potential for over diagnosis of higher functioning autism and Asperger syndrome. However, the etiology of ASDs is largely unknown, and few genetic tests or biomarkers have been found to confirm an autism diagnosis. In other words, no laboratory-based means of testing for autism is yet widely available. While various genes have been identified that cause autism, none of them are common, and justification for performing individual testing is debatable. However, tests are currently being offered commercially for a number of these known genes, including *CNTNAP2*, *PTEN* and *SHANK3* . Recently, microarray technology has also been proposed as a potential diagnostic tool, or a method to determine etiology. Because of the growing number of genes known to cause ASDs, an autism microarray, containing only verified autism causing genes, could be custom made to assess if these genes are aberrant in affected individuals. The proportion of cases with known etiology at this stage will be very small (less than 5%) and a negative test would not mean an exclusion of the diagnosis. Knowing the root cause of autism may not lead to an alteration in treatment or intervention at this stage, but may be useful for family planning and genetic counseling, as well as for the emotional well being of concerned family members. Additionally, with further phenotype profiling and analysis of individuals with particular mutations, it may be possible to tailor interventions based on genetic diagnoses in the future.

2. Causes of ID

ID can be caused by environmental and/or genetic factors. However, for up to 60% of cases, there is no identifiable cause (Rauch et al., 2006). Environmental exposure to certain teratogens, viruses or radiation can cause ID, as can severe head trauma or injury causing lack of oxygen to the brain (Chelly et al., 2006). While these factors explain some cases of ID, it is also important to consider genetic etiology.

Genetic causes of ID are thought to be present in 25-50% of cases, although this number increases proportionally with severity (McLaren & Bryson, 1987). Chromosomal abnormalities have been reported in ID, with a broad range of prevalence, and many different types of aberrations have been identified (Rauch et al., 2006). Over the past 15 years many single gene causes of ID have been identified as well. Many of these genes cause a broad range of phenotypes including syndromic ID (S-ID), non-syndromic ID (NS-ID), autism and other neurodevelopmental and psychiatric phenotypes. This suggests that it is likely that other genetic modifiers or environmental factors may be involved in disease etiology, and that similar biological pathways, when disturbed, have the potential to lead to a range of these conditions. This illustrates the need for detailed study and descriptions of phenotypes for each gene and mutation. Most of the mutations that cause ID are highly penetrant and are inherited in a Mendelian fashion. Many known ID genes are on the X-chromosome, however the number of autosomal genes associated with ID is growing rapidly (Kaufman et al., 2010). This is due to advances in technology which allow us to study the autosomes more efficiently, along with a shift in focus from the X-chromosome to the autosomes resulting from the realization that the X-linked genes may only account for ~10% of ID cases, while autosomal genes may account for many more (Ropers & Hamel, 2005).

3. Causes of autism

3.1 Genetic contributions

The causes of autism are likely far more complex than the causes of ID. Genetic factors clearly play a prominent role. The evidence for the involvement of genetic factors in the etiology of autism comes primarily from family and twin studies and is further supported by cytogenetic and molecular studies. Several sibling concordance studies have provided a strong indication that autism has a significant genetic component. Studies over the years, including analyses of autism probands with severe ID, evaluations of the developmental, social, and psychiatric histories of siblings of autistic individuals and reviews of familial aggregation in large autism cohorts, have all displayed higher than expected familiality (Baird & August, 1985; Piven et al., 1990; Bolton et al., 1994). A recent study examined the cognitive, adaptive, social, imitation, play, and language abilities of 42 non-autistic siblings (of autism probands) and 20 toddlers with no family history of autism. The siblings were below average in expressive language abilities and IQ. They had lower mean receptive language, adaptive behavior, and social communication skills. They used fewer words, distal gestures, and social smiles than children with no familial history of autism (Toth et al., 2007).

On the other hand, the familiality of autism does not imply that genetic factors are exclusively responsible for the disease, and role of the environmental factors, which are also shared by family members who live together, cannot be excluded by these studies alone. Twin studies provide an alternative approach for investigating the relative magnitude of genetic and/or environmental factors on the autism phenotype and penetrance. In 1977, a landmark study by Folstein and Rutter demonstrated a significant difference between monozygotic (MZ) and dizygotic (DZ) twins in their concordance for autism (Folstein & Rutter, 1977). This difference in concordance suggested a major role for genes in the etiology of autism and this was confirmed by subsequent studies (Ritvo et al., 1985; Steffenburg et al., 1989). Most recently, a large scale study of 277 twin pairs (210 DZ and 67 MZ) reported 88% concordance of autism for MZ twins and 31% concordance of autism for DZ twins. In MZ twins, the authors also observed a higher prevalence of bipolar disorder and Asperger syndrome with a higher concordance of the latter (Rosenberg et al., 2009). The finding of increased bipolar disorder in twins is interesting because the co-morbidity of psychiatric disorders (eg. anxiety disorder, ADHD, etc) in children with autism is up to 70% (Charman et al., 2011), and suggests significant overlap in genetic etiology between different psychiatric disorders. Specific genetic causes for autism will be outlined more thoroughly later in this chapter.

3.2 Epigenetic contributions

It is widely accepted that genetic factors play a major role in the etiology of ASD, however, epigenetic factors may also be an important determinant of the autism phenotype. Epigenetic modifications include cytosine methylation and post-translational modification of histone proteins, and act as a mechanism to control gene expression (Samaco et al., 2005). The epigenetic regulation of gene expression can be influenced by exposure to environmental factors and can show parent of origin effects. Notably, epigenetic factors play a central role in pathogenesis of two single gene disorders, Rett syndrome and Fragile X syndrome (FXS), that are commonly associated with autism (Brown et al., 1982; Gillberg, 1986). Rett syndrome, a progressive neurodevelopmental disorder, is classified among

ASDs. It is caused by mutation in the *MECP2* gene that encodes the methyl-CpG-binding protein 2, which is involved in epigenetic regulation of gene expression (Amir et al., 1999). *JARID1C*, a relatively common ID gene associated with some cases of autism (Adegbola et al., 2008) has been shown to regulate the methylation of histones and function in epigenetic transcriptional repression (Tahiliani et al., 2007).

In addition, some autism (and ID) genes are regulated by known epigenetic mechanisms. FXS is caused by the expansion of a tract of CGG repeats in the 5′ untranslated region of the *FMR1* gene. This expansion results in epigenetic silencing of the region, causing loss of expression of the gene, thus, FXS is caused by a genetic mutation resulting in epigenetic dysregulation (Hagerman et al., 2005). The *RELN* gene is another interesting example of possible contributions of epigenetic factors to ASD. The *RELN* gene encodes a large extracellular matrix protein that organizes neuronal positioning during corticogenesis and is regulated by epigenetic mechanisms. Several independent studies have shown an association between *RELN* and ASD (Skaar et al., 2005; Ashley-Koch et al., 2007; Holt et al., 2010). Interestingly, reduced levels of reelin and its isoforms have been previously shown in autistic twins and their first degree relatives (Fatemi et al., 2005).

Genomic imprinting is another mode of regulation of gene expression by epigenetic modifications and it results in parent of origin-specific gene expression. Genomic duplications of an imprinted region on the proximal long arm of chromosome 15 (15q11-q13) are associated with 0.5-3.0% of autism cases (Hogart et al., 2010).

3.3 Other factors

Support for possible environmental factors contributing to the causation of autism comes from the incomplete concordance in monozygotic twins. Additionally, at this point in time, known genetic defects only explain a small proportion of autism patients. Furthermore, there is evidence that *in utero* exposure to valproic acid and thalidomide may increase the risk of ASD (Arndt et al., 2005). One long-term study of 632 children exposed to antiepileptic drugs during gestation, found that 6.3% of the children exposed to valproic acid *in utero* had ASD or some features of ASD. This incidence is seven times higher than the control group (0.9%; (Bromley et al., 2008)). Similarly, a higher incidence of autism has been reported among children prenatally exposed to thalidomide. In a population of 100 Swedish thalidomide embryopathy cases, at least four met full diagnostic criteria for autism (Stromland et al., 1994). Animal models have also demonstrated that early serotonergic neural development is disrupted in rats exposed to thalidomide or valproic acid on the ninth day of gestation, conferring increased risk for the development of ASD-related behaviours (Narita et al., 2010).

Mercury (Hg), because of its known neurotoxicity, has drawn particular attention in relation to the neurodevelopment of individuals with autism, and a number of studies have compared the level of Hg in blood, hair, or urine in children with autism versus without autism. However, none of these studies have shown any substantial evidence of involvement of Hg in autism. Recently, a study conducted on 452 autism patients failed to demonstrate any difference in blood Hg level of autism patients compared to controls (Hertz-Picciotto et al., 2010).

Childhood immunization is an environmental factor that has been popularized in the media as a potential cause of autism. The use of mercury in vaccines has been one of the prime sources of concern surrounding vaccines and their role in autism (Baker, 2008). However, there is no consistent evidence in support of the theory that vaccines are related to the etiology of autism. In the late 1990s, a link between vaccines and autism was reported by clinical observation of the onset of autism soon after vaccination of children (Wakefield et al., 1998). These observations triggered a series of studies in the US, UK, Europe and Japan, however, none of these studies found any compelling evidence for a link between vaccines and autism. The original study has since been retracted (Murch et al., 2004; Anonymous, 2010)

Although the majority of research to date has focused on genetic factors involved in the etiology of autism, non-genetic factors are also likely to contribute. Our knowledge of these factors is, however, currently very limited. It has been suggested that distinct genetic features/pathways may cause distinct domains of autistic behaviour, but this has yet to be tested at the molecular level (Happe & Ronald, 2008). It does, however, resonate with the idea that autism is a genetically heterogeneous spectrum, and that multiple genetic aberrations may be necessary to reach the autism phenotype threshold (Cook & Scherer, 2008). The threshold theory postulates that the cumulative effect of several genetic aberrations, for instance a copy number variant together with one or more single nucleotide variants, and possibly in combination with environmental factors, in a single individual, may result in an autism phenotype. These genetic aberrations may include chromosomal, single nucleotide or epigenetic abnormalities. It has also been noted that some genetic aberrations are more penetrant than others and may be more likely to result in a phenotype. In contrast with ID genetics, which are relatively straightforward, autism presents us with a convoluted, likely multigenic/multifactorial disorder for which it may be more difficult to delineate causes.

4. Epidemiology

4.1 Autism & autism spectrum disorders

In published literature the incidence of autism is variable, and a worldwide trend of increase in prevalence of autism has been reported. During the 1980s, autism was thought to be rare, with a prevalence of less than 5 per 10,000 persons (Gillberg et al., 1991) and was not categorized as major public health problem. During the 1990s, the prevalence of autism was estimated to be 21 to 31 per 10,000 in preschool children (Fombonne, 1999). A recent review of epidemiologic studies reported a prevalence of 20 per 10,000 for classic autism and 60-70 per 10,000 (1 in 150) for all ASDs (Fombonne, 2009). In addition it reported a prevalence of 30-40 per 10,000 for PDD-NOS and 2 per 10,000 for CDD (Fombonne, 2009). Epidemiologic studies of Asperger syndrome have been more rare, and although current numbers estimate a prevalence of 6 per 10,000, there are severe limitations to calculating this prevalence accurately (Fombonne, 2009). Other recent large scale epidemiological studies have shown that as many as 1 in 100, or 1% of school age children have an ASD (Baird et al., 2006; Baron-Cohen et al., 2009; Kogan et al., 2009). It is noteworthy that some of these studies are based on parents' reporting of ASD and it could be argued that these estimates might be falsely high (Kogan et al., 2009). On the other hand, it has been argued that the increasing incidence of autism might be due to increased awareness of public and professionals coupled with the broadening of the diagnostic criteria (Fombonne et al., 2006). Today, the prevalence of ASDs is believed to be very high and this condition is now thought to be second only to Intellectual Disability (ID) among the most common developmental disabilities in the United States (Yeargin-Allsopp et al., 2003).

It is also noteworthy that there is a gender bias in autism. Among children with autism, the ratio of affected males to females is estimated to be 4:1 (Volkmar et al., 1993) and the male to female ratio for Asperger syndrome is even higher. In contrast, Rett syndrome occurs almost exclusively in females.

4.2 Intellectual disabilities

ID is the most common neurodevelopmental disorder in the United States (Yeargin-Allsopp, et al., 2003). The prevalence of ID is between 1 and 3% (Roeleveld et al., 1997; Leonard & Wen, 2002) and is present in every social class and culture (Leonard & Wen, 2002). Despite its universal occurrence, there tends to be higher prevalence of ID in areas of lower socioeconomic status and developing countries, particularly for mild cases (Drews et al., 1995; Roeleveld et al., 1997; Durkin et al., 1998; Durkin, 2002; Emerson, 2007). The variability of prevalence is more pronounced for mild ID than for severe forms. It has been suggested that this discrepancy is likely due to environmental factors (Roeleveld et al., 1997; Durkin et al., 1998; Emerson, 2007).

Approximately 30% more males are diagnosed with ID than females (McLaren & Bryson, 1987; American Psychiatric Association, 2000). However, despite a higher ratio of males to females among milder cases of ID, the ratio decreases as IQ decreases (McLaren & Bryson, 1987; American Psychiatric Association, 2000). Some studies suggest that severe ID may be more prevalent among females (Katusic et al., 1996; Bradley et al., 2002), however these studies were performed in quite specific geographic locations and populations, and may not necessarily be generalizable to other regions. Some of this gender bias can be accounted for by mutations on the X-chromosome. In most cases of X-linked ID (ie. X-linked mental retardation; XLMR) or X-linked autism, more males are affected due to hemizygosity. However, in some disorders, such as Rett syndrome, this ratio is reversed because mutations in the Rett syndrome gene, *MECP2*, are generally lethal in haploid genomes. In addition, a rare phenomenon is apparent in female-restricted epilepsy and mental retardation (EFMR), in which heterozygous mutations in the gene *PCDH19* cause the disease in females and in which there is reprieve-in-males with hemizygous *PCDH19* mutations, who remain unaffected (Dibbens et al., 2008; Hynes et al., 2009). A possible explanation for this reprieve-in-males phenomenon could be that carrier males have a homogenous population of mutant *PCDH19*-containing cells, whereas affected females would possess a mosaic population of mutant and wild-type *PCDH19*-containing cells. It is postulated that this mosaicism, rather than the effect of the mutated protein alone, may disrupt cell-cell communication, resulting in the clinical presentation (Dibbens et al., 2008).

4.3 Co-morbidity of autism spectrum disorders & intellectual disability (syndromic and non-syndromic)

It is often necessary to look at ASDs and ID together, as there is significant overlap between them both in terms of phenotype and in genetic causation. As previously noted, ID is present in ~50-60% of individuals with autism. Additionally, in a study performed on an ID population, 28% met the criteria for an autism diagnosis on the ADI-R scale and only half of these cases had been previously diagnosed (Bryson et al., 2008). Similar studies have also shown that within ID populations, the prevalence of autism is 8-20%, and that it is more likely for individuals with severe ID to meet criteria for ASD (Wing & Gould, 1979; Deb &

Prasad, 1994; Nordin & Gillberg, 1996; Stromme & Diseth, 2000; de Bildt et al., 2005; Bryson et al., 2008).

ID and autism have multiple overlapping phenotypic domains. The three major phenotypic domains that characterize autism—language deficits, social deficits and stereotypies/ repetitive behaviours—can often be seen to varying degrees in individuals with ID. Individuals with ID often display stereotypies, which tend to become more pronounced and often self-injurious with decreasing IQ (Symons et al., 2005). Studies have found that 30-60% of individuals with ID display some form of stereotypy (Bodfish et al., 1995; Bodfish et al., 2000; Goldman et al., 2009). Language deficits are often particularly pronounced in individuals with severe and profound ID.

Many ID syndromes have an occurrence of autism that is significantly higher than the occurrence for the general population. For example, a current review of the literature indicates that up to 25-47% of individuals with Fragile X syndrome, 5-10% of individuals with Down syndrome, and 16-48% of individuals with tuberous sclerosis (TSC) also have an autism/PDD diagnosis, compared to 0.6-1% in the general population (Fombonne, 2009; Molloy et al., 2009). Other ID syndromes that have high occurrence of concordant autism include Angelman syndrome, Joubert Syndrome and Cohen syndrome.

5. Shared genetics of autism and ID

5.1 Shared genetic causes of autism and ID: Syndromic

As previously noted, there is evidence that several syndromic forms of ID are more likely to present with autism than would be expected in the general population. The observation of overlap in phenotype between autism and the most common XLMR disorder, fragile X mental retardation syndrome (FXS) is a long-standing, albeit controversial one. The frequency of molecular diagnosis of FXS among autistic patients has been reported as high as 12.4% and as low as 0%, averaging at 7.25% (Smalley et al., 1988; Gurling et al., 1997). More recent studies indicate a more conservative rate of FXS of 2-4% (Wassink et al., 2001). Two studies of young FXS individuals demonstrated that 25% and 33% met criteria for autism and a review suggests that as many as 47% may have an ASD (Bailey et al., 1998; Rogers et al., 2001; Molloy et al., 2009); however, it has also been argued that autistic features are not more common among individuals with FXS than among other individuals with ID (Bardoni et al., 2000). Mutations in *MECP2* have also been found among individuals with autism (Kim & Cook, 2000; Orrico et al., 2000; Beyer et al., 2002; Hammer et al., 2002). Autistic features are also frequently present in other ID syndromes such as Down syndrome and phenylketonuria (PKU).

These disorders are rarely mistaken for autism, as other syndromic features assist with the correct diagnostic assignment. On the other hand, there is evidence that even for these well-characterized syndromic forms of XLMR, there is a very broad phenotypic expression of the disease. For instance, mutations within the aristaless gene, *ARX*, are responsible for several distinct forms of XLMR and neurological phenotypes: West syndrome (infantile spasms with hypsarrhythmia; Stromme et al. 2002a), Partington syndrome (XLMR with dystonic hand movements; Stromme et al. 2002b), XLAG (Kitamura et al. 2002), XLMR (Bienvenu et al. 2002), Proud syndrome (Kato et al. 2004), and various forms of epilepsy (Stromme et al. 2002a, b; Scheffer et al. 2002). A 24 base pair duplication in *ARX* has been found in families with West syndrome and Partington syndrome, and recently in several families previously identified as having non-syndromic XLMR. In these families several individuals were reported as having only mild intellectual impairment along with autism or autistic-like behaviours (Turner et al., 2002).

5.2 Shared genetic causes of autism and ID: Non-syndromic

The Neuroligin 4 (*NLGN4*) gene has been linked to autism by several studies (Jamain et al., 2003; Laumonnier et al., 2004; Marshall et al., 2008). However, in 2004, Laumonnier et al. identified a family containing individuals with NS-ID, with or without ASD, segregating with a *NLGN4* mutation (Laumonnier et al., 2004). This study was the first to suggest that NS-ID and autism may have overlapping genetic etiologies. A similar finding was noted with *NRXN1*, which interacts with *NLGN4*. Heterozygous copy number variants (CNV) in this gene have been found in autism while homozygous mutations of *NRXN1* cause the S-ID disorder Pitt-Hopkins-like syndrome-2 (PTHSL2; Zweier et al., 2009).

More recently, a truncating mutation was found in *SHANK3* in an individual with NS-ID (Hamdan et al., 2011). *SHANK3* has been found to cause autism in several studies (Durand et al., 2007; Marshall et al., 2008). In addition, *SHANK2* CNVs and sequencing mutations have been found in several cases of both autism and NS-ID, displaying a significant level of etiological overlap between the two disorders (Berkel et al., 2010; Pinto et al., 2010). CNVs, or structural variation within the genome, appear to contribute significantly to the etiology of ID and autism.

PTCHD1 is another X-linked gene that has been implicated in autism and NS-ID. A CNV which deletes *PTCHD1* entirely causes NS-ID in one family (Noor et al., 2010). Another CNV, which results in a loss of the first exon and upstream region of *PTCHD1*, results in autism in another family (Noor et al., 2010). In addition, one CNV upstream of the gene was found in an individual with ADHD, suggesting that it may play a role in this phenotype as well (Noor et al., 2010). *IL1RAPL1*, which was initially identified as a cause of NS-ID, and has been shown to cause NS-ID in several individuals, has also been implicated in autism (Carrie et al., 1999; Bhat et al., 2008; Marshall et al., 2008; Piton et al., 2008; Pinto et al., 2010). Similarly, a missense mutation in the NS-ID gene *JARID1C* was found in an individual with autism (Adegbola et al., 2008). Most recently, a *de novo* CNV deletion overlapping *SYNGAP1*, a gene previously implicated in NS-ID, was identified in a female autism proband (Pinto et al., 2010). These genetic links are of much interest, particularly due to the strong phenotypic overlap seen in NS-ID and autism. These common genes will be an important factor in teasing out which biochemical processes are disturbed in different forms of developmental delay, and why a particular mutation in an individual may lead to one condition rather than the other.

6. Common biological pathways (see Table 2)

6.1 The mTOR pathway (see Figure 1)

The common pathways shared by autism genes are also particularly interesting with respect to ID. While it cannot necessarily be argued that all autism genes will fall into distinct and neat categories, there are certain pathways which are overrepresented among autism-related genes identified so far. One review of the literature suggested that there appear to be two major pathways that known autism genes are a part of: 1. excitation and inhibition at the synaptic junction and 2. cellular and synaptic growth—i.e. participation in the mTOR pathway (Bourgeron, 2009). This categorization is valid but does not encompass all autism genes, and as our knowledge of genes that contribute to the autism phenotype increases, more common pathways may be elucidated.

Gene Name	ASD or ID*	Chromosomal Locus	Protein Product	Potential Pathogenic Mechanism
Common Gene Function: Cell Adhesion				
CDH8	ASD	16q22.1	Cadherin 8	Disruption of neuronal cell adhesion leading to aberrant synaptogenesis or plasticity
CDH9	ASD	5p14	Cadherin 9	As above
CDH10	ASD	5p14.2	Cadherin 10	As above
CDH15	ID	16q24.3	Cadherin 15	As above
CNTNAP2	ASD	7q35	Contactin-associated protein-like 2	As above
NLGN3	ASD	Xq13.1	Neuroligin 3	Disruption of neuronal cell adhesion specifically at the excitatory synapse leading to aberrant synaptogenesis or plasticity
NLGN4	ASD/ID	Xp22.33	Neuroligin 4	As above
NRXN1	ASD/ID	2p16.3	Neurexin 1	Disruption of neuronal cell adhesion leading to aberrant synaptogenesis or plasticity
PCDH10	ASD	4q28.3	Protocadherin 10	As above
PCDH19	ID	Xq13.3	Protocadherin 19	As above
TSPAN7	ASD/ID			As above
Common Gene Function: Receptors at Inhibitory Synapse				
GABA receptors (Chr 4)	ASD	4p14	GABA receptors α2, β1, γ1 and α4	Disruption of GABA receptor activity at the inhibitory synapse
GABA receptors (Chr 15)	ASD/ID	15q11.2-q12	GABA receptors α5, β3, and γ3	Disruption of GABA receptor activity at the inhibitory synapse
Common Gene Function: Regulation and Organization at the Excitatory Synapse				
FMR1	ASD/ID	Xq27.3	Fragile X mental retardation 1 protein	Aberrant translational regulation of important synaptic genes
GRIK2	ASD/ID	6q16.3-q21	Glutamate receptor, ionotropic, kainate 2	Disruption of Kainate Receptors at the excitatory synapse

Gene Name	ASD or ID*	Chromosomal Locus	Protein Product	Potential Pathogenic Mechanism
IL1RAPL1	ASD/ID	Xp22.1-p21.3	Interleukin 1 receptor accessory protein-like 1	Disruption of activity and organization at the excitatory synapse
SHANK2	ASD/ID	1q41	SH3 and multiple ankyrin repeat domains 2	As Above
SHANK3	ASD/ID	22q13.3	SH3 and multiple ankyrin repeat domains 3	As Above
SYNGAP1	ASD/ID	6p21.3	Synaptic Ras GTPase activating protein 1	Disruption of NMDA and AMPA receptors via down regulation of Ras/ERK signalling
Common Gene Function: Transcriptional Control				
ARX	ASD/ID	Xp21.3	Aristaless related homeobox	Disruption of transcriptional regulation leading to alterations in dosage of multiple genes
AUTS2	ASD	7q11.2	Autism susceptibility gene 2 protein	Neuronal nuclear expression; Putative regulator of transcription (Kalscheuer et al., 2007)
JARID1C	ASD/ID	Xp11.22-p11.21	Jumonji, AT rich interactive domain 1C	Disruption of transcriptional regulation leading to alterations in dosage of multiple genes
MECP2	ASD/ID	Xq28	Methyl CpG binding protein 2	As above
Common Gene Function: Down Regulation of the mTOR Signaling Pathway				
NF1	ASD/ID	17q11.2	neurofibromin	Neuronal cell overgrowth due to the down regulation of the mTOR pathway or down regulation of Ras/ERK signaling
PTEN	ASD	10q23.3	phosphatidylinositol -3,4,5-trisphosphate 3-phosphatase and dual-specificity protein phosphatase	Neuronal cell overgrowth due to the down regulation of the mTOR pathway
TSC1	ASD/ID	9q34	Tuberous Sclerosis 1 protein (hamartin)	As above

Gene Name	ASD or ID*	Chromosomal Locus	Protein Product	Potential Pathogenic Mechanism
TSC2	ASD/ID	16p13.3	Tuberous Sclerosis 2 protein (tuberin)	As above
Common Gene Function: Varying Functionalities				
IQSEC2	ASD/ID	Xp11.22	IQ motif and SEC7 domain-containing protein 2	Disruption of the ARF signaling pathway
PTCHD1	ASD/ID	Xp22.11	Patched domain 1	Potential disruption of the Hedgehog signaling pathway
RELN	ASD/ID	7q22	Reelin	Abnormal neuronal cell migration and cell-cell interaction
SLC6A8	ASD/ID	Xq28	Solute carrier family 6 member 8	Creatine deficiency
UPF3B	ASD/ID	Xq25-q26	UPF3 regulator of nonsense transcripts homolog B	Dysfunction of nonsense mediated decay and mRNA surveillance

*associated with autism spectrum disorder or intellectual disability

Table 2. Genes associated with ASD, ID or both and chromosomal location, along with protein product and function, and the potential route by which the gene results in neurodevelopmental phenotypes.

NF1, *TSC1* and *TSC2* are genes that, when mutated, may result in disorders - neurofibromatosis and tuberous sclerosis respectively- in which there is high incidence of autism- neurofibromatosis and tuberous sclerosis respectively. These genes are negative regulators of the mTOR-raptor complex (mTORC1)—a complex that is important in regulation and cell growth during mitosis and likely plays an active role in synaptogenesis (Bourgeron, 2009). Furthermore, in hippocampus of fmr1 knockout mice, upregulation of mTOR activity has also been reported (Sharma et al., 2010). It can be speculated that when deleterious mutations occur in these genes, the mTOR pathway would become more active due to loss of down regulation. *PTEN*, another negative regulator of the mTOR pathway has also been implicated in autism. *PTEN* mutations can lead to PTEN Hamartoma-Tumor Syndrome (PHTS) which includes Cowden Syndrome and Bannayan-Riley-Ruvalcaba Syndrome, but are also present in autism probands without these syndromes (McBride et al., 2010). The mutations found in autism probands appear to be less penetrant than other *PTEN* mutations (McBride et al., 2010).

With the finding that mTOR genes are involved in autism susceptibility we can propose several possibilities, the first of which is that regulators of the mTORC1 complex may also be good candidates for ID genes, based on overlapping phenotype between autism and ID. However, individuals with neurofibromatosis do not typically present with ID despite having a higher incidence of learning disabilities, ADHD and autism (Hsueh, 2007). Tuberous Sclerosis is associated with learning disabilities, developmental delay including autism and epilepsy in about 85% of cases (Curatolo et al., 2008). This specificity for autism susceptibility in the mTOR pathway may help us to delineate the complex association between autism and ID genes, and understand how autism specific phenotypes are caused.

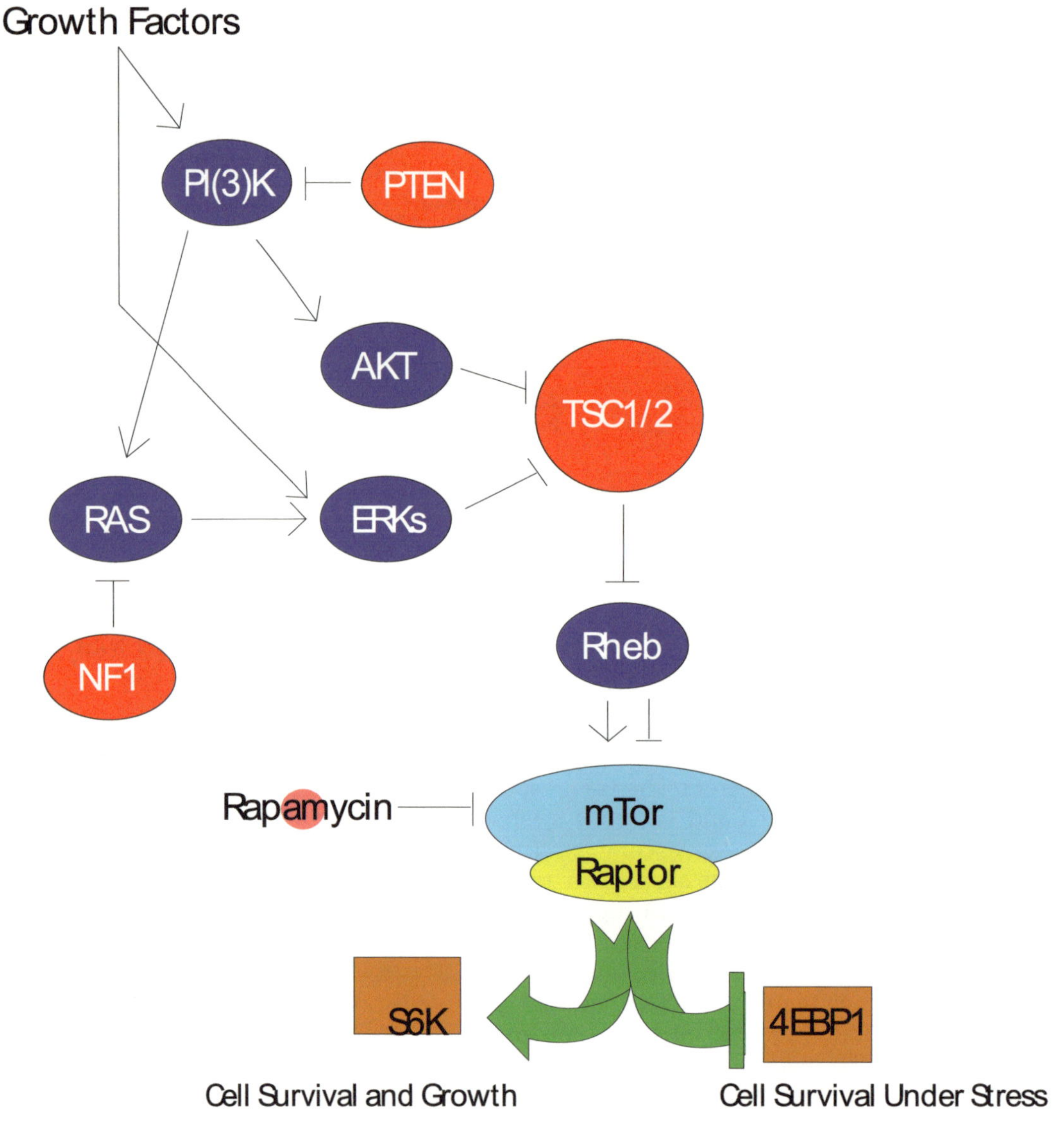

The mTOR pathway has been implicated as a potential major contributor to the etiology of autism. The mTOR pathway is negatively regulated by NF1, TSC1/2, and PTEN (and possibly FMRP), all of which are involved in the etiology of autism. mTOR activation results in cell survival and proliferation. Negative regulation of this pathway prevents overgrowth. Some cancer syndromes are also associated with these genes. mTOR is a target of rapamycin, which inhibits its functioning. This is a potential pharmaceutical candidate for the treatment of syndromes caused by mutations in these genes, as well as autism resulting from mutations in these genes. This figure shows the role of NF1, TSC1/2 and PTEN in the mTOR pathway.

Fig. 1. mTor Pathway in autism and ID

6.2 Synaptic proteins and synaptogenesis (see Figures 2 & 3)

Most of the known genes that overlap both autism and ID are present at the synapse and are involved in the excitatory/inhibitory pathway, with an emphasis on excitation. For example, the SHANK family of genes appears to make quite a significant contribution to autism and ID etiologies. Both *SHANK2* and *SHANK3* encode scaffolding proteins present at the post-synaptic density and in dendrites. They are important for scaffolding in the post-synaptic density—connecting ion channels, neurotransmitter receptors and other membrane proteins to the actin cytoskeleton—and act as a structural framework at this site (Boeckers et al., 2002; Hayashi et al., 2009). They are also likely to play a role in neuronal plasticity (Boeckers et al., 2002; Hayashi et al., 2009).

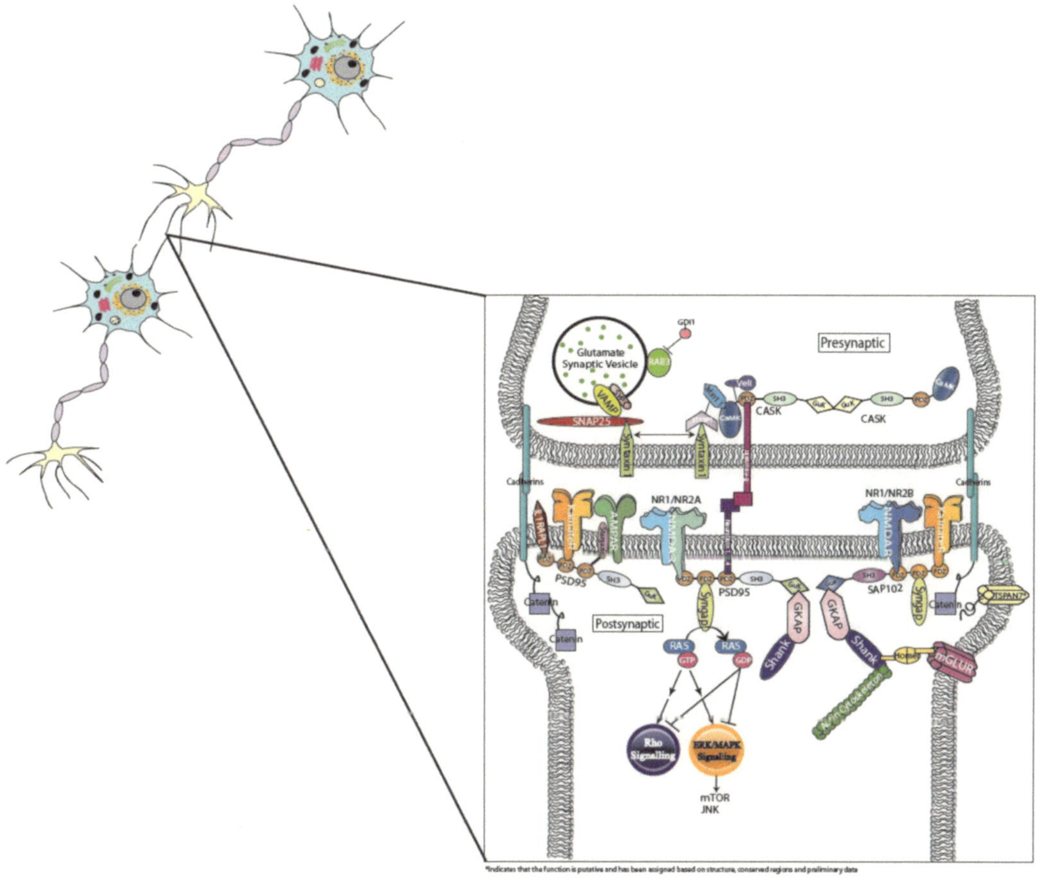

Many autism genes, in particular those that overlap with intellectual disabilities, are present at excitatory synapses. Many of these genes encode protein products which are present in the postsynaptic density (PSD) including *SYNGAP1, SHANK2, SHANK3, NLGN4, NLGN3, NRXN1,* and *IL1RAPL1*. Glutamatergic synapses contain NMDA receptors, AMPA receptors and Kainate receptors. Mutations in additional synaptic genes have been implicated in ID, e.g. *CASK* and *STXBP1*. Aberrant function at this synapse has been postulated to be part of both autism and ID etiology.

Fig. 2. Glutamatergic Synapse in autism and ID

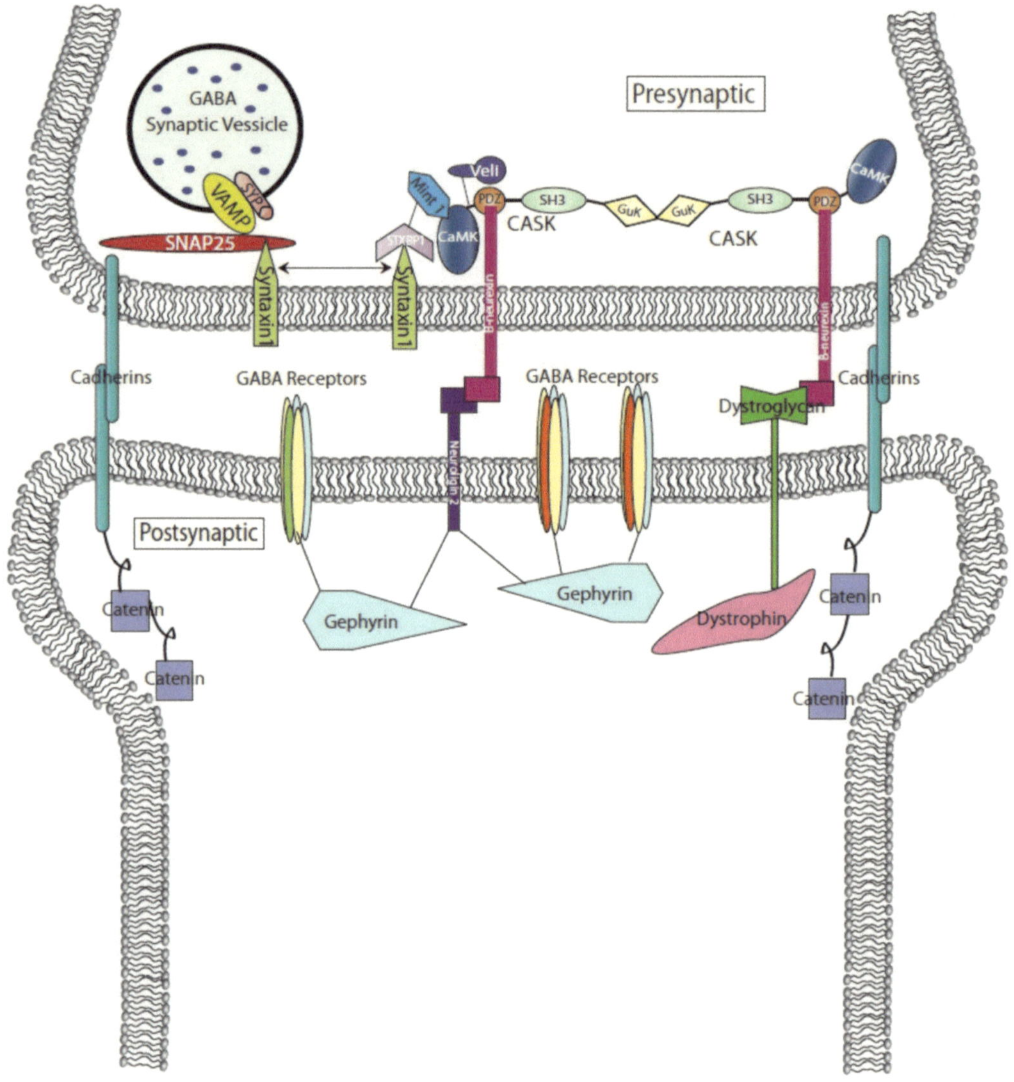

GABA is the major inhibitory neurotransmitter in the human brain. GABA receptor genes have been postulated as candidates for autism etiology. Most GABA receptors cluster in certain chromosomal regions some of which are associated with autism and/or ID, such as 15q11.2-q13 which contains GABRα5, GABRβ3, and GABRγ3, and 4p14 which contains the genes GABRα2, GABRβ1, GABRγ1 and GABRα4. Duplications of the 15q11.2-q13 region are present in 0.5-3% of autism cases.

Fig. 3. GABAergic Synapse in autism and ID

Another important gene involved in both ID and autism is *SYNGAP1* which has been identified as a dominant cause of ID in several individuals with truncating mutations (Hamdan et al, 2009; Hamdan et al, 2011), as well a cause of autism in an individual with a CNV deletion overlapping the gene (Pinto et al. 2010). *SYNGAP1* encodes SynGAP—a GTPase activating protein that is part of the NMDA receptor complex (NMDAR), and binds to the NR2B subunit (Kim et al., 2005). NMDARs play a role in glutamate-activated excitation of postsynaptic neurons, and have been implicated in memory formation and synaptic plasticity. SynGAP is a negative regulator of NMDAR mediated ERK activation and causes inhibition of the Ras/ERK pathway (Kim et al., 2005). Over-expression of SynGAP has also been shown to down regulate GLuR1, a subunit of AMPA receptors (AMPAR), a class of excitatory ionotropic glutamate receptors which are regulated by the Ras/ERK pathway (Kim et al., 2005; Rumbaugh et al., 2006). *Syngap* knockout mice implicate SynGAP in the regulation of long term potentiation (LTP) and AMPAR expression (Komiyama et al., 2002). While both ID and autism probands with mutations in this gene were heterozygous for the mutation, individuals with ID have truncating single nucleotide mutations, while the individual with autism had a deleterious CNV, which is an example of how different aberrations within the same gene can have differential effects on phenotypic manifestation, i.e. allelism (Hamdan et al., 2009; Pinto et al., 2010; Hamdan et al., 2011). This has been shown for several other autism/ID genes as well, including *IL1RAPL1*, *JARID1C*, and *SHANK2* (Adegbola et al., 2008; Piton et al., 2008; Berkel et al., 2010; Pinto et al., 2010).

Other autism-related genes also bind to the NMDAR and are involved in its function. *NF1* described earlier as a down regulator of the mTOR pathway also plays a role in the negative regulation of the Ras signaling pathway and is known to bind directly to the NMDAR complex (Hsueh, 2007). This suggests an alternate mechanism for its role in autism. Within the context of Bourgeron's review, this gene is actually affiliated with both "major" autism gene pathways.

In addition *IL1RAPL1* also appears to have an impact on NMDAR function. Mutations in *IL1RAPL1*, a gene with several known mutations in NS-ID and autism, result in the incorrect localization of the MAGUK family protein PSD-95 (*DLG4*), which is important for organization and function of NMDARs, ion channels and other signaling proteins at the post-synapse (Carrie et al., 1999; Gardoni, 2008; Pavlowsky et al., 2010). The IL1RAPL1 protein has been shown to interact with PSD-95, and knockout of this gene decreased the post-synaptic density (PSD) and the localization of PSD-95 at excitatory synapses. Loss of *IL1RAPL1* also results in a decrease of activity in the JNK pathway, which was found to lead to decreased phosphorylation of PSD-95 (Pavlowsky et al., 2010). *IL1RAPL1* has also been shown to be important for the formation of excitatory synapses *in vivo* (Pavlowsky et al., 2010). PSD-95 directly interacts with several known NS-ID associated proteins including: CASK, SynGAP, GLuR6 and neuroligins (Kim & Sheng, 2004).

GRIK2 is a gene that is mutated in NS-ID (Motazacker et al., 2007) and has been linked by association studies to autism (Jamain et al., 2002; Shuang et al., 2004; Kim et al., 2007) . It encodes a protein called GLuR6, which is a subunit of a kainate receptor (KAR). KARs are ionotropic glutamate receptors which respond to the excitatory neurotransmitter l-glutamate, similar to NMDA or AMPA receptors. They are expressed at a high level in the brain, particularly in the hippocampal mossy fibers, where GLuR6 has been found to modulate LTP in mouse models (Bortolotto et al., 1999; Contractor, Swanson and Heinemann., 2001). GLuR6 knockout mice show decreased LTP in mossy fibers. LTP in the hippocampus has been implicated as a mechanism for memory formation and learning (Bliss & Collingridge, 1993; Fedulov et al., 2007). While polymorphisms in *GRIK2* are likely not responsible for an autism phenotype alone, the association suggests that this locus confers susceptibility to the phenotype and may contribute to the disorder in concert with other genetic aberrations and environmental factors.

Additionally, FMRP, encoded for by *FMR1*, the fragile X syndrome (FXS) gene, is thought to play a role in neuronal plasticity by acting as a suppressor of local protein translation (as reviewed by Jin et al. 2004). It binds mRNA in the nucleus and carries the mRNA to its target destination in the cytoplasm (Bagni & Greenough, 2005). The transcripts targeted by FMRP include some that are relevant to autism and ID, including *PSD-95*, *DLG1*, *SHANK1*, *DLGAP1-4*, *Grin1*, *Grin2b*, *GluR1*, and *GluR2* (Bassell & Warren, 2008; Schutt et al., 2009). It also binds other important neuronal transcripts such as itself (*FMR1*), *SEMA3F*, *CamKIIα*, *GABRD*, *ARC*, *MAP1B* and *APP* (Bassell & Warren, 2008). While FXS is an X-linked syndrome of variable phenotype, it is a very common cause of ID, and often is present together with autistic behaviours (de Vries et al., 1998). In addition, the *FMR2* gene, which is similar to *FMR1* in structure but has a poorly defined function, is also involved in the genetics of ID and autism (Gecz et al., 1996). Mutations in this gene result in intellectual disabilities with or without autistic behaviours.

Some important synaptic genes that are present at inhibitory synaptic junctions appear to be associated with autism and some ID syndromes. GABA is the major inhibitory neurotransmitter in the human brain, and dysfunction of its receptors could result in a decrease of the inhibitory response. A significant amount of research has been done to study the role of GABA and GABA receptors in autism. In particular, GABA receptor subunits residing in clusters on 4p and 15q have been implicated in autism. As indicated earlier, genomic duplications of an imprinted region on the long arm of chromosome 15 (15q11-q13) are present in 0.5-3.0% of autistic individuals (Hogart et al., 2010). Deletions and uniparental disomy (UPD) in this region are responsible for Angelman syndrome and Prader-Willi syndrome depending on the parent of origin of the mutation, and both of these syndromes present with ID (Dykens et al., 2004), and frequently also with autism (Hogart et al., 2010). This region contains a cluster of three GABA receptor subunit genes: *GABRa5*, *GABRβ3*, and *GABRγ3* (Dykens et al., 2004). Based on the frequency of this duplication in autism, it is possible that these genes are involved in the autism phenotype, however there are ~10 genes in the duplicated region, all of which may play a major, minor or no role in the autism phenotype. *GABRa2*, *GABRβ1*, *GABRγ1* and *GABRa4* on 4p14 (Ma et al., 2005; Vincent et al., 2006; Kakinuma et al., 2008) have also been implicated in autism. Molecular work specific to some of these GABA receptors supports their role in autism (Ma et al., 2005; Collins et al., 2006; Fatemi et al., 2010). It is also of note that GABA neurotransmission is strongly implicated in fragile X syndrome, and knockout of the *fmr1* gene in mice has a hugely disruptive effect on the GABAergic system, and is a potential target for the treatment of symptoms for both Fragile-X syndrome and autism (Hagerman et al., 2005).

Several studies have found significant genetic associations between the chromosome 4 GABA receptor cluster and autism. In addition, specific GABA receptor genes within the 4p cluster have been implicated as likely contributors to autism. *GABRa4* was found to be involved in the etiology of autism independently and through interactions with *GABRβ1* (Ma et al., 2005) and both of these genes have been linked to the autism phenotype by association (Collins et al., 2006). Recently, a study by Fatemi et al. indicated that the levels of GABA receptor mRNAs in autism brains are significantly different from controls (Fatemi et al., 2010). The study shows that in the BA9 region of brains acquired from individuals with autism, levels of *GABRa4*, *GABRa5* and *GABRβ1* mRNAs are significantly decreased, while in the cerebella of these brains mRNA for these same genes are increased compared to normal controls after normalization with housekeeping genes (Fatemi et al., 2010). In addition, several small studies and cases have shown that the levels of GABA in peripheral blood and plasma are altered in individuals with autism (Dhossche et al., 2002, Rolf et al., 1993) but these findings are inconsistent, and more thorough and larger scale studies need to be done to determine whether GABA could act as stable biomarker for ASD.

6.3 Neuronal cell adhesion

Neuronal cell adhesion is an interesting common pathway between ID and autism. As previously noted, *NLGN4* and *NRXN1* are both examples of genes that are mutated in cases of autism and ID. *NLGN4* presents us with a particularly interesting genetic link between autism and ID, as it displayed pleiotropy within a single family, with the same mutation causing autism in some individuals, and NS-ID in others. The NLGN4 protein, located on the postsynaptic membrane, interacts with NRXN1 on the presynaptic membrane. Heterozygous mutations in *NRXN1* appear to result in autism as well (Autism Genome Project Consortium et al., 2007; Kim et al., 2008). Similar disruptions in *NRXN1* have also been documented in schizophrenia (Rujescu et al., 2009), ID, and language delays (Ching et al., 2010). The NLGN4 protein acts as an important element in postsynaptic differentiation, forming complexes with β-neurexins and PSD-95 (Ichtchenko et al., 1995; Irie et al., 1997; Scheiffele et al., 2000). NLGN4 is linked to glutamatergic postsynaptic proteins and neuroligin/neurexin complexes, which appears to be sufficient for synaptogenesis (Graf et al., 2009). These genes play a major role in both cell adhesion as well as synaptogenesis showing a role for overlap between these pathways.

Additionally, it has been postulated that *TSPAN7*, an X-linked NS-ID gene, is involved in a complex of ß-integrins, which are involved in cell-cell and cell-matrix interactions (Zemni et al., 2000) and that this gene may cause autism in individuals with deleterious CNVs (Marshall et al., 2008). There are several other neuronal cell adhesion genes that have been implicated in autism, with and without ID, including *NLGN3* and *CNTNAP2* (Jamain et al., 2003; Alarcon et al., 2008). This suggests that neuronal cell adhesion is a common mechanism by which both autism and ID occur, and may be helpful in elucidating the biological mechanism of these highly related, albeit different, disorders.

In addition to these neuronal cell adhesion genes, several cadherins and protocadherins have been implicated in autism as well. *CDH8* has been found to be disrupted by microdeletions in individuals with learning disabilities and autism, and is not disrupted in over 5000 controls (Pagnamenta et al., 2011). Additionally, a genome-wide association study of individuals with autism identified significant peaks at *CDH9* and *CDH10* (Wang et al., 2009). *PCDH10* has also been suggested as a candidate gene for autism based on a homozygous CNV deletion overlapping the gene in an affected individual (Morrow et al., 2008).

Several similar genes have been implicated in ID as well. *PCDH19*, a protocadherin, has also been implicated in epilepsy with mental retardation limited to females (EFMR; Dibbens et al., 2008; Hynes et al., 2009). Additionally *CDH15* is an autosomal dominant cause of S-ID and NS-ID in several individuals. *CDH15* encodes a cadherin that is expressed mainly in brain and skeletal muscle (Bhalla et al., 2008). Mutations of this gene in individuals with ID were found to decrease cell adhesion by greater than 80% (Bhalla et al., 2008). It is clear from these genetic links that neuronal cell adhesion is a common pathway in both ID and autism. Interestingly, while some of these genes overlap both ID and ASD, some are unique to one condition or the other. It is possible that some of these apparently unique genes may actually be involved in both disorders but no examples have been identified because mutations are so rare. It is also possible that these genes could only cause specific endophenotypes and may be useful for helping us tease out the intricate web of connections between the two disorders.

6.4 Transcriptional control

Disruption of transcriptional control can have far reaching implications in terms of phenotypic manifestation. Genes involved in transcriptional control, when disrupted, may in turn affect the expression of many other genes. For example, *ARX* is one of the most frequently mutated genes in X-linked NS-ID. It is a homeobox-containing gene that is part of the Aristaless-related gene family. This is a family of transcription factors that are required for various essential events during vertebrate embryogenesis, including CNS development (Meijlink et al. 1999). Based on experimental data and gene structure it has been speculated that *ARX* regulates transcription by both gene activation and suppression and it is essential for normal development of the CNS (as reviewed by Friocourt et al. 2006). Mutations in this gene have been implicated in many ID syndromes, XLMR and autism/autistic features (Turner et al., 2002; Friocourt et al., 2006).

Another example of a transcription factor involved in both autism and ID is *MECP2*, the causative gene for Rett syndrome: a regressive syndrome described earlier in this chapter. *MECP2* mutations may result in various alternative phenotypic manifestations including MRXS13, LUBS X-linked ID syndrome and NS-ID. While genotype/phenotype data is often ambiguous, some studies have demonstrated a genotype/phenotype correlation in terms of severity, as well as for specific phenotypic measures (Ham et al., 2005; Bebbington et al., 2008). *MECP2* encodes the methyl CpG binding protein 2 (MECP2), which is believed to act as a transcriptional modulator that is capable of long-range chromatin re-organization resulting in repression or activation of genes through binding to methylated CpG DNA (As reviewed by Gonzales and LaSalle, 2010).

Additionally, *JARIDIC*, also known as *KMD5C*, is another relatively common gene related to X-linked ID with over twenty mutations known in XLMR individuals (Tzschach et al., 2006; Tahiliani et al., 2007). In addition, it has been identified as a causative gene in an individual with autism (Adegbola et al., 2008). *JARIDIC* is a histone demethylase containing a PHD-finger domain that is characteristic of zinc finger proteins and specifically demethylates di- and tri-methylated histone 3 lysine 4 (H3K4me2/me3) residues (Christensen et al., 2007; Tahiliani et al., 2007; Cloos et al., 2008). Trimethylation at this residue is extremely important for transcriptional regulation and chromatin structure. *JARID1C* is likely involved in repressor element silencing transcription factor (REST)-mediated transcriptional repression. It has been shown to regulate the expression of several REST-mediated genes, as well as regulate the H3K4me2/H3K4me3 levels at their promoters (Tahiliani et al., 2007). Determining which genes these transcriptional regulators control may give insight into the biological pathways involved in disease and the mechanisms by which they occur.

The gene *AUTS2* was originally identified through the mapping of a translocation breakpoint on chromosome 7 in a pair of autistic monozygotic twins (Sultana et al., 2002), and subsequently identified in a number of studies of other autism patients with cytogenetic aberrations (Bakkaloglu et al., 2008; Huang et al., 2010). *AUTS2* has more recently been identified at the breakpoint for *de novo* translocations in three unrelated ID individuals (Kalscheuer et al., 2007), as well as at breakpoints of CNVs in individuals with ADHD (Elia et al., 2010) and epilepsy (Mefford et al.. 2010). Although the function of the protein encoded by *AUTS2* is unknown, *in silico* analysis suggest the protein has similarity to known transcription factors (Kalscheuer et al., 2007), and expression studies show that the protein is highly expressed within the nucleus of neurons and neuronal progenitors during development of the cerebral cortex and cerebellum, as well as other regions (Bedogni et al., 2010). As such, it is likely that AUTS2 also functions as a regulator of gene transcription.

7. Other genes displaying ID/autism overlap (see Table 2)

While it is very helpful to look at overlapping autism and ID genes in common pathways, it is not always possible. Autism and ID both have highly variable genetic causes. Some of these causal genes cannot be grouped together in common pathways but are still very important and have been widely implicated in autism and ID. *PTCHD1* is an example of such a gene. This gene is estimated to explain ~1% of cases of autism, as well as several cases of NS-ID (Noor et al., 2010). *PTCHD1* is thought to encode a receptor for the hedgehog signaling pathway, however no definitive role for the PTCHD1 protein has yet been established, and as the patched-like domain present in *PTCHD1* also has a potential sterol-sensing function, there may be sterol transporting pathways implicated (Noor et al., 2010). Mutations at the *PTCHD1* locus can occur either within the gene itself or in the region upstream of the gene, which are thought to disrupt *PTCHD1* regulation. Additionally, *UPF3B*, a component of the nonsense mediated decay surveillance machinery, has been implicated in various ID cases across four families, as well as in several individuals with autism (Tarpey et al., 2007; Addington et al., 2010; Laumonnier et al., 2010). It has also been implicated in childhood onset schizophrenia and ADHD (Addington et al., 2010). *IQSEC2*, a gene that encodes a GTPase for the ARF family of proteins, is also mutated in several families with ID and varying degrees of ASD and epilepsy (Shoubridge et al., 2010).

8. Summary

In this chapter, we have explored the relationship between ASD and ID: two separate but often co-morbid forms of developmental disorder. They are both relatively common in the general population, however the incidence of ASD appears to be on the rise, while ID is relatively stable. Both can be impacted by environmental factors, however ASD appears to have a more complex etiology and may require a combination of genetic, epigenetic and environmental factors to manifest phenotypically. Meanwhile, genes that cause ID tend to be either *de novo* or passed down in a Mendelian fashion, and are highly penetrant.

While ASD is co-morbid with ID in 40-70% of cases, ASD can also present with normal intelligence. Certain endophenotypes, such as IQ, head size, presence of dysmorphisms, seizures and MRI abnormalities, may help to predict the effectiveness of early intensive behavioural intervention. Understanding the fundamental differences between "essential" and "complex" autism may be the key to creating personalized behavioural programming that is specific not only for the skills of a particular child, but for the phenotypic specificity conferred by a narrower diagnosis.

It is clear that many of the genes implicated in both ASD and ID cause variable developmental, intellectual and psychiatric phenotypes, with and without additional clinical symptoms. Understanding the molecular mechanisms that result in developmental delays will be useful for potential management and interventions in these disorders. Understanding more about the phenotypes conferred by aberrations in different genes may lead us to develop differential interventions based on these genotypes. Additionally, classification of aberrant molecular pathways may help us to identify biomarkers, which could be used for early diagnosis of these disorders. Individuals with ASD classically respond best to early intensive intervention, thus the earlier we can diagnose ASD, the earlier we can act to help ensure the best outcomes.

9. References

Anonymous (2010) Retraction--Ileal-Lymphoid-Nodular Hyperplasia, Non-Specific Colitis, and Pervasive Developmental Disorder in Children. Lancet 375:445

Abrahams BS and Geschwind DH (2008) Advances in Autism Genetics: On the Threshold of a New Neurobiology. Nat Rev Genet 9:341-355

Addington AM, Gauthier J, Piton A, et al (2010) A Novel Frameshift Mutation in UPF3B Identified in Brothers Affected with Childhood Onset Schizophrenia and Autism Spectrum Disorders. Mol Psychiatry

Adegbola A, Gao H, Sommer S, Browning M (2008) A Novel Mutation in JARID1C/SMCX in a Patient with Autism Spectrum Disorder (ASD). Am J Med Genet A 146A:505-511

Alarcon M, Abrahams BS, Stone JL, et al (2008) Linkage, Association, and Gene-Expression Analyses Identify CNTNAP2 as an Autism-Susceptibility Gene. Am J Hum Genet 82:150-159

American Psychiatric Association (2000) Diagnostic and Statistical Manual of Mental Disorders, Fourth Edition, Text Revision. American Psychiatric Association, Washington

Amir RE, Van den Veyver IB, Wan M, Tran CQ, Francke U, Zoghbi HY (1999) Rett Syndrome is Caused by Mutations in X-Linked MECP2, Encoding Methyl-CpG-Binding Protein 2. Nat Genet 23:185-188

Arndt TL, Stodgell CJ, Rodier PM (2005) The Teratology of Autism. Int J Dev Neurosci 23:189-199

Ashley-Koch AE, Jaworski J, Ma de Q, et al (2007) Investigation of Potential Gene-Gene Interactions between APOE and RELN Contributing to Autism Risk. Psychiatr Genet 17:221-226

Autism Genome Project Consortium, Szatmari P, Paterson AD, et al (2007) Mapping Autism Risk Loci using Genetic Linkage and Chromosomal Rearrangements. Nat Genet 39:319-328

Bagni C and Greenough WT (2005) From mRNP Trafficking to Spine Dysmorphogenesis: The Roots of Fragile X Syndrome. Nat Rev Neurosci 6:376-387

Bailey DB,Jr, Mesibov GB, Hatton DD, Clark RD, Roberts JE, Mayhew L (1998) Autistic Behavior in Young Boys with Fragile X Syndrome. J Autism Dev Disord 28:499-508

Baird G, Simonoff E, Pickles A, Chandler S, Loucas T, Meldrum D, Charman T (2006) Prevalence of Disorders of the Autism Spectrum in a Population Cohort of Children in South Thames: The Special Needs and Autism Project (SNAP). Lancet 368:210-215

Baird TD and August GJ (1985) Familial Heterogeneity in Infantile Autism. J Autism Dev Disord 15:315-321

Baker JP (2008) Mercury, Vaccines, and Autism: One Controversy, Three Histories. Am J Public Health 98:244-253

Bakkaloglu B, O'Roak BJ, Louvi A, et al (2008) Molecular cytogenetic analysis and resequencing of contactin associated protein-like 2 in autism spectrum disorders. Am J Hum Genet 82:165-173

Bardoni B, Mandel JL, Fisch GS (2000) FMR1 Gene and Fragile X Syndrome. Am J Med Genet 97:153-163

Baron-Cohen S, Scott FJ, Allison C, Williams J, Bolton P, Matthews FE, Brayne C (2009) Prevalence of Autism-Spectrum Conditions: UK School-Based Population Study. Br J Psychiatry 194:500-509

Bassell GJ and Warren ST (2008) Fragile X Syndrome: Loss of Local mRNA Regulation Alters Synaptic Development and Function. Neuron 60:201-214

Bebbington A, Anderson A, Ravine D, et al (2008) Investigating Genotype-Phenotype Relationships in Rett Syndrome using an International Data Set. Neurology 70:868-875

Bedogni F, Hodge RD, Nelson BR, Frederick EA, Shiba N, Daza RA, Hevner RF (2010) Autism susceptibility candidate 2 (Auts2) encodes a nuclear protein expressed in developing brain regions implicated in autism neuropathology. Gene Expr Patterns 10:9-15.

Berkel S, Marshall CR, Weiss B, et al (2010) Mutations in the SHANK2 Synaptic Scaffolding Gene in Autism Spectrum Disorder and Mental Retardation. Nat Genet

Bertrand J, Mars A, Boyle C, Bove F, Yeargin-Allsopp M, Decoufle P (2001) Prevalence of Autism in a United States Population: The Brick Township, New Jersey, Investigation. Pediatrics 108:1155-1161

Beyer KS, Blasi F, Bacchelli E, Klauck SM, Maestrini E, Poustka A, International Molecular Genetic Study of Autism Consortium (IMGSAC) (2002) Mutation Analysis of the Coding Sequence of the MECP2 Gene in Infantile Autism. Hum Genet 111:305-309

Bhalla K, Luo Y, Buchan T, et al (2008) Alterations in CDH15 and KIRREL3 in Patients with Mild to Severe Intellectual Disability. Am J Hum Genet 83:703-713

Bhat SS, Ladd S, Grass F, Spence JE, Brasington CK, Simensen RJ, Schwartz CE, Dupont BR, Stevenson RE, Srivastava AK (2008) Disruption of the IL1RAPL1 Gene Associated with a Pericentromeric Inversion of the X Chromosome in a Patient with Mental Retardation and Autism. Clin Genet 73:94-96

Bliss TV and Collingridge GL (1993) A Synaptic Model of Memory: Long-Term Potentiation in the Hippocampus. Nature 361:31-39

Boddaert N, Zilbovicius M, Philipe A, et al (2009) MRI Findings in 77 Children with Non-Syndromic Autistic Disorder. PLoS One 4:e4415

Bodfish JW, Crawford TW, Powell SB, Parker DE, Golden RN, Lewis MH (1995) Compulsions in Adults with Mental Retardation: Prevalence, Phenomenology, and Comorbidity with Stereotypy and Self-Injury. Am J Ment Retard 100:183-192

Bodfish JW, Symons FJ, Parker DE, Lewis MH (2000) Varieties of Repetitive Behavior in Autism: Comparisons to Mental Retardation. J Autism Dev Disord 30:237-243

Boeckers TM, Bockmann J, Kreutz MR, Gundelfinger ED (2002) ProSAP/Shank Proteins - a Family of Higher Order Organizing Molecules of the Postsynaptic Density with an Emerging Role in Human Neurological Disease. J Neurochem 81:903-910

Bolton P, Macdonald H, Pickles A, Rios P, Goode S, Crowson M, Bailey A, Rutter M (1994) A Case-Control Family History Study of Autism. J Child Psychol Psychiatry 35:877-900

Bortolotto ZA, Clarke VR, Delany CM, et al (1999) Kainate Receptors are Involved in Synaptic Plasticity. Nature 402:297-301 Bourgeron T (2009) A Synaptic Trek to Autism. Curr Opin Neurobiol 19:231-234

Bradley EA, Thompson A, Bryson SE (2002) Mental Retardation in Teenagers: Prevalence Data from the Niagara Region, Ontario. Can J Psychiatry 47:652-659

Bromley RL, Mawer G, Clayton-Smith J, Baker GA, Liverpool and Manchester Neurodevelopment Group (2008) Autism Spectrum Disorders Following in Utero Exposure to Antiepileptic Drugs. Neurology 71:1923-1924

Brown WT, Friedman E, Jenkins EC, Brooks J, Wisniewski K, Raguthu S, French JH (1982) Association of Fragile X Syndrome with Autism. Lancet 1:100

Bryson SE, Bradley EA, Thompson A, Wainwright A (2008) Prevalence of Autism among Adolescents with Intellectual Disabilities. Can J Psychiatry 53:449-459

Carrie A, Jun L, Bienvenu T, et al (1999) A New Member of the IL-1 Receptor Family Highly Expressed in Hippocampus and Involved in X-Linked Mental Retardation. Nat Genet 23:25-31

Chakrabarti S and Fombonne E (2001) Pervasive Developmental Disorders in Preschool Children. JAMA 285:3093-3099

Charman T, Jones CR, Pickles A, Simonoff E, Baird G, Happe F (2011) Defining the Cognitive Phenotype of Autism. Brain Res 1380:10-21

Chelly J, Khelfaoui M, Francis F, Cherif B, Bienvenu T (2006) Genetics and Pathophysiology of Mental Retardation. Eur J Hum Genet 14:701-713

Ching MS, Shen Y, Tan WH, et al (2010) Deletions of NRXN1 (Neurexin-1) Predispose to a Wide Spectrum of Developmental Disorders. Am J Med Genet B Neuropsychiatr Genet 153B:937-947

Christensen J, Agger K, Cloos PA, Pasini D, Rose S, Sennels L, Rappsilber J, Hansen KH, Salcini AE, Helin K (2007) RBP2 Belongs to a Family of Demethylases, Specific for Tri-and Dimethylated Lysine 4 on Histone 3. Cell 128:1063-1076

Clarke DF, Roberts W, Daraksan M, Dupuis A, McCabe J, Wood H, Snead OC,3rd, Weiss SK (2005) The Prevalence of Autistic Spectrum Disorder in Children Surveyed in a Tertiary Care Epilepsy Clinic. Epilepsia 46:1970-1977

Cloos PA, Christensen J, Agger K, Helin K (2008) Erasing the Methyl Mark: Histone Demethylases at the Center of Cellular Differentiation and Disease. Genes Dev 22:1115-1140

Collins AL, Ma D, Whitehead PL, et al (2006) Investigation of Autism and GABA Receptor Subunit Genes in Multiple Ethnic Groups. Neurogenetics 7:167-174

Contractor A, Swanson G, Heinemann SF (2001) Kainate Receptors are Involved in Short- and Long-Term Plasticity at Mossy Fiber Synapses in the Hippocampus. Neuron 29:209-216

Cook EH,Jr and Scherer SW (2008) Copy-Number Variations Associated with Neuropsychiatric Conditions. Nature 455:919-923

Curatolo P, Bombardieri R, Jozwiak S (2008) Tuberous Sclerosis. Lancet 372:657-668 de Bildt A, Sytema S, Kraijer D, Minderaa R (2005) Prevalence of Pervasive Developmental Disorders in Children and Adolescents with Mental Retardation. J Child Psychol Psychiatry 46:275-286

de Vries BB, Halley DJ, Oostra BA, Niermeijer MF (1998) The Fragile X Syndrome. J Med Genet 35:579-589

Deb S and Prasad KB (1994) The Prevalence of Autistic Disorder among Children with a Learning Disability. Br J Psychiatry 165:395-399

Dhossche D, Applegate H, Abraham A, Maertens P, Bland L, Bencsath A, Martinez J (2002) Elevated Plasma Gamma-Aminobutyric Acid (GABA) Levels in Autistic Youngsters: Stimulus for a GABA Hypothesis of Autism. Med Sci Monit 8:PR1-6

Dibbens LM, Tarpey PS, Hynes K, et al (2008) X-Linked Protocadherin 19 Mutations Cause Female-Limited Epilepsy and Cognitive Impairment. Nat Genet 40:776-781

Drews CD, Yeargin-Allsopp M, Decoufle P, Murphy CC (1995) Variation in the Influence of Selected Sociodemographic Risk Factors for Mental Retardation. Am J Public Health 85:329-334

DSM-5 Proposed Revisions Include New Category of Autism Spectrum Disorders (2010) American Psychiatric Association. Release No. 10-09. February 10 2010

Durand CM, Betancur C, Boeckers TM, et al (2007) Mutations in the Gene Encoding the Synaptic Scaffolding Protein SHANK3 are Associated with Autism Spectrum Disorders. Nat Genet 39:25-27

Durkin M (2002) The Epidemiology of Developmental Disabilities in Low-Income Countries. Ment Retard Dev Disabil Res Rev 8:206-211

Durkin MS, Hasan ZM, Hasan KZ (1998) Prevalence and Correlates of Mental Retardation among Children in Karachi, Pakistan. Am J Epidemiol 147:281-288

Dykens EM, Sutcliffe JS, Levitt P (2004) Autism and 15q11-q13 Disorders: Behavioral, Genetic, and Pathophysiological Issues. Ment Retard Dev Disabil Res Rev 10:284-291

Elia J, Gai X, Xie HM, et al. (2010) Rare structural variants found in attention-deficit hyperactivity disorder are preferentially associated with neurodevelopmental genes. Mol Psychiatry 15:637-646

Emerson E (2007) Poverty and People with Intellectual Disabilities. Ment Retard Dev Disabil Res Rev 13:107-113

Fatemi SH, Snow AV, Stary JM, Araghi-Niknam M, Reutiman TJ, Lee S, Brooks AI, Pearce DA (2005) Reelin Signaling is Impaired in Autism. Biol Psychiatry 57:777-787

Fatemi SH, Reutiman TJ, Folsom TD, Rooney RJ, Patel DH, Thuras PD (2010) MRNA and Protein Levels for GABAAalpha4, alpha5, beta1 and GABABR1 Receptors are Altered in Brains from Subjects with Autism. J Autism Dev Disord 40:743-750

Fedulov V, Rex CS, Simmons DA, Palmer L, Gall CM, Lynch G (2007) Evidence that Long-Term Potentiation Occurs within Individual Hippocampal Synapses during Learning. J Neurosci 27:8031-8039

Folstein S and Rutter M (1977) Infantile Autism: A Genetic Study of 21 Twin Pairs. J Child Psychol Psychiatry 18:297-321

Fombonne E (1999) The Epidemiology of Autism: A Review. Psychol Med 29:769-786

Fombonne E (2003) Epidemiological Surveys of Autism and Other Pervasive Developmental Disorders: An Update. J Autism Dev Disord 33:365-382

Fombonne E, Zakarian R, Bennett A, Meng L, McLean-Heywood D (2006) Pervasive Developmental Disorders in Montreal, Quebec, Canada: Prevalence and Links with Immunizations. Pediatrics 118:e139-50

Fombonne E (2009) Epidemiology of Pervasive Developmental Disorders. Pediatr Res 65:591-598

Friocourt G, Poirier K, Rakic S, Parnavelas JG, Chelly J (2006) The Role of ARX in Cortical Development. Eur J Neurosci 23:869-876

Gardoni F (2008) MAGUK Proteins: New Targets for Pharmacological Intervention in the Glutamatergic Synapse. Eur J Pharmacol 585:147-152

Gecz J, Gedeon AK, Sutherland GR, Mulley JC (1996) Identification of the Gene FMR2, Associated with FRAXE Mental Retardation. Nat Genet 13:105-108

Gillberg C (1986) Autism and Rett Syndrome: Some Notes on Differential Diagnosis. Am J Med Genet Suppl 1:127-131

Gillberg C, Steffenburg S, Schaumann H (1991) Is Autism More Common Now than Ten Years Ago? Br J Psychiatry 158:403-409

Goldman S, Wang C, Salgado MW, Greene PE, Kim M, Rapin I (2009) Motor Stereotypies in Children with Autism and Other Developmental Disorders. Dev Med Child Neurol 51:30-38

Gonzales ML and LaSalle JM (2010) The Role of MeCP2 in Brain Development and Neurodevelopmental Disorders. Curr Psychiatry Rep 12:127-134

Graf ER, Daniels RW, Burgess RW, Schwarz TL, DiAntonio A (2009) Rab3 Dynamically Controls Protein Composition at Active Zones. Neuron 64:663-677

Gurling HM, Bolton PF, Vincent J, Melmer G, Rutter M (1997) Molecular and Cytogenetic Investigations of the Fragile X Region Including the Frax A and Frax E CGG Trinucleotide Repeat Sequences in Families Multiplex for Autism and Related Phenotypes. Hum Hered 47:254-262

Hagberg B, Aicardi J, Dias K, Ramos O (1983) A Progressive Syndrome of Autism, Dementia, Ataxia, and Loss of Purposeful Hand use in Girls: Rett's Syndrome: Report of 35 Cases. Ann Neurol 14:471-479

Hagerman RJ, Ono MY, Hagerman PJ (2005) Recent Advances in Fragile X: A Model for Autism and Neurodegeneration. Curr Opin Psychiatry 18:490-496

Ham AL, Kumar A, Deeter R, Schanen NC (2005) Does Genotype Predict Phenotype in Rett Syndrome? J Child Neurol 20:768-778

Hamdan FF, Gauthier J, Spiegelman D, et al (2009) Mutations in SYNGAP1 in Autosomal Nonsyndromic Mental Retardation. N Engl J Med 360:599-605

Hamdan FF, Gauthier J, Araki Y, et al (2011) Excess of *De Novo* Deleterious Mutations in Genes Associated with Glutamatergic Systems in Nonsyndromic Intellectual Disability. Am J Hum Genet 88:306-316

Hammer S, Dorrani N, Dragich J, Kudo S, Schanen C (2002) The Phenotypic Consequences of MECP2 Mutations Extend Beyond Rett Syndrome. Ment Retard Dev Disabil Res Rev 8:94-98

Happe F and Ronald A (2008) The 'Fractionable Autism Triad': A Review of Evidence from Behavioural, Genetic, Cognitive and Neural Research. Neuropsychol Rev 18:287-304

Hara H (2007) Autism and Epilepsy: A Retrospective Follow-Up Study. Brain Dev 29:486-490

Hayashi MK, Tang C, Verpelli C, Narayanan R, Stearns MH, Xu RM, Li H, Sala C, Hayashi Y (2009) The Postsynaptic Density Proteins Homer and Shank Form a Polymeric Network Structure. Cell 137:159-171

Hertz-Picciotto I, Green PG, Delwiche L, Hansen R, Walker C, Pessah IN (2010) Blood Mercury Concentrations in CHARGE Study Children with and without Autism. Environ Health Perspect 118:161-166

Hogart A, Wu D, LaSalle JM, Schanen NC (2010) The Comorbidity of Autism with the Genomic Disorders of Chromosome 15q11.2-q13. Neurobiol Dis 38:181-191

Holt R, Barnby G, Maestrini E, et al (2010) Linkage and Candidate Gene Studies of Autism Spectrum Disorders in European Populations. Eur J Hum Genet 18:1013-1019

Howlin P, Goode S, Hutton J, Rutter M (2004) Adult Outcome for Children with Autism. J Child Psychol Psychiatry 45:212-229

Hsueh YP (2007) Neurofibromin Signaling and Synapses. J Biomed Sci 14:461-466

Huang XL, Zou YS, Maher TA, Newton S, Milunsky JM (2010) A *de novo* balanced translocation breakpoint truncating the autism susceptibility candidate 2 (AUTS2) gene in a patient with autism. Am J Med Genet A 152A:2112-2114.

Hughes JR and Melyn M (2005) EEG and Seizures in Autistic Children and Adolescents: Further Findings with Therapeutic Implications. Clin EEG Neurosci 36:15-20

Hurley RS, Losh M, Parlier M, Reznick JS, Piven J (2007) The Broad Autism Phenotype Questionnaire. J Autism Dev Disord 37:1679-1690

Hynes K, Tarpey P, Dibbens LM, et al (2009) Epilepsy and Mental Retardation Limited to Females with PCDH19 Mutations can Present *De Novo* Or in Single Generation Families. J Med Genet

Ichtchenko K, Hata Y, Nguyen T, Ullrich B, Missler M, Moomaw C, Sudhof TC (1995) Neuroligin 1: A Splice Site-Specific Ligand for Beta-Neurexins. Cell 81:435-443

Irie M, Hata Y, Takeuchi M, Ichtchenko K, Toyoda A, Hirao K, Takai Y, Rosahl TW, Sudhof TC (1997) Binding of Neuroligins to PSD-95. Science 277:1511-1515

Jamain S, Betancur C, Quach H, Philippe A, Fellous M, Giros B, Gillberg C, Leboyer M, Bourgeron T, Paris Autism Research International Sibpair (PARIS) Study (2002) Linkage and Association of the Glutamate Receptor 6 Gene with Autism. Mol Psychiatry 7:302-310

Jamain S, Quach H, Betancur C, et al (2003) Mutations of the X-Linked Genes Encoding Neuroligins NLGN3 and NLGN4 are Associated with Autism. Nat Genet 34:27-29

Jin P, Alisch RS, Warren ST (2004) RNA and microRNAs in Fragile X Mental Retardation. Nat Cell Biol 6:1048-1053

Kakinuma H, Ozaki M, Sato H, Takahashi H (2008) Variation in GABA-A Subunit Gene Copy Number in an Autistic Patient with Mosaic 4 p Duplication (p12p16). Am J Med Genet B Neuropsychiatr Genet 147B:973-975

Kalscheuer VM, FitzPatrick D, Tommerup N, et al (2007) Mutations in Autism Susceptibility Candidate 2 (AUTS2) in Patients with Mental Retardation. Hum Genet 121:501-509

Kanner L (1943) Autistic Disturbances of Affective Contact. Nervous Child 2: 217-250

Katusic SK, Colligan RC, Beard CM, O'Fallon WM, Bergstralh EJ, Jacobsen SJ, Kurland LT (1996) Mental Retardation in a Birth Cohort, 1976-1980, Rochester, Minnesota. Am J Ment Retard 100:335-344

Kaufman L, Ayub M, Vincent JB (2010) The Genetic Basis of Non-Syndromic Intellectual Disability: A Review. J Neurodev Disord 2:182-209

Kim SJ and Cook EH,Jr (2000) Novel *De Novo* Nonsense Mutation of MECP2 in a Patient with Rett Syndrome. Hum Mutat 15:382-383

Kim E and Sheng M (2004) PDZ Domain Proteins of Synapses. Nat Rev Neurosci 5:771-781

Kim MJ, Dunah AW, Wang YT, Sheng M (2005) Differential Roles of NR2A- and NR2B-Containing NMDA Receptors in Ras-ERK Signaling and AMPA Receptor Trafficking. Neuron 46:745-760

Kim SA, Kim JH, Park M, Cho IH, Yoo HJ (2007) Family-Based Association Study between GRIK2 Polymorphisms and Autism Spectrum Disorders in the Korean Trios. Neurosci Res 58:332-335

Kim HG, Kishikawa S, Higgins AW, et al (2008) Disruption of Neurexin 1 Associated with Autism Spectrum Disorder. Am J Hum Genet 82:199-207

Kleinman JM, Robins DL, Ventola PE, et al (2008) The Modified Checklist for Autism in Toddlers: A Follow-Up Study Investigating the Early Detection of Autism Spectrum Disorders. J Autism Dev Disord 38:827-839

Kogan MD, Blumberg SJ, Schieve LA, Boyle CA, Perrin JM, Ghandour RM, Singh GK, Strickland BB, Trevathan E, van Dyck PC (2009) Prevalence of Parent-Reported Diagnosis of Autism Spectrum Disorder among Children in the US, 2007. Pediatrics 124:1395-1403

Komiyama NH, Watabe AM, Carlisle HJ, et al (2002) SynGAP Regulates ERK/MAPK Signaling, Synaptic Plasticity, and Learning in the Complex with Postsynaptic Density 95 and NMDA Receptor. J Neurosci 22:9721-9732

Laumonnier F, Bonnet-Brilhault F, Gomot M, et al (2004) X-Linked Mental Retardation and Autism are Associated with a Mutation in the NLGN4 Gene, a Member of the Neuroligin Family. Am J Hum Genet 74:552-557

Laumonnier F, Shoubridge C, Antar C, et al (2010) Mutations of the UPF3B Gene, which Encodes a Protein Widely Expressed in Neurons, are Associated with Nonspecific Mental Retardation with Or without Autism. Mol Psychiatry 15:767-776

Leonard H and Wen X (2002) The Epidemiology of Mental Retardation: Challenges and Opportunities in the New Millennium. Ment Retard Dev Disabil Res Rev 8:117-134

Lord C, Rutter M, Le Couteur A (1994) Autism Diagnostic Interview-Revised: A Revised Version of a Diagnostic Interview for Caregivers of Individuals with Possible Pervasive Developmental Disorders. J Autism Dev Disord 24:659-685

Lord C, Risi S, Lambrecht L, Cook EH,Jr, Leventhal BL, DiLavore PC, Pickles A, Rutter M (2000) The Autism Diagnostic Observation Schedule-Generic: A Standard Measure of Social and Communication Deficits Associated with the Spectrum of Autism. J Autism Dev Disord 30:205-223

Ma DQ, Whitehead PL, Menold MM, et al (2005) Identification of Significant Association and Gene-Gene Interaction of GABA Receptor Subunit Genes in Autism. Am J Hum Genet 77:377-388

Marshall CR, Noor A, Vincent JB, et al (2008) Structural Variation of Chromosomes in Autism Spectrum Disorder. Am J Hum Genet 82:477-488

Maulik PK, Mascarenhas MN, Mathers CD, Dua T, Saxena S (2011) Prevalence of Intellectual Disability: A Meta-Analysis of Population-Based Studies. Res Dev Disabil 32:419-436

McBride KL, Varga EA, Pastore MT, Prior TW, Manickam K, Atkin JF, Herman GE (2010) Confirmation Study of PTEN Mutations among Individuals with Autism Or Developmental delays/mental Retardation and Macrocephaly. Autism Res 3:137-141

McConachie H and Diggle T (2007) Parent Implemented Early Intervention for Young Children with Autism Spectrum Disorder: A Systematic Review. J Eval Clin Pract 13:120-129

McLaren J and Bryson SE (1987) Review of Recent Epidemiological Studies of Mental Retardation: Prevalence, Associated Disorders, and Etiology. Am J Ment Retard 92:243-254

Mefford HC, Muhle H, Ostertag P, et al (2010) Genome-wide copy number variation in epilepsy: novel susceptibility loci in idiopathic generalized and focal epilepsies. PLoS Genet 6:e1000962

Miles JH, Hadden LL, Takahashi TN, Hillman RE (2000) Head Circumference is an Independent Clinical Finding Associated with Autism. Am J Med Genet 95:339-350

Miles JH and Hillman RE (2000) Value of a Clinical Morphology Examination in Autism. Am J Med Genet 91:245-253

Miles JH, Takahashi TN, Bagby S, Sahota PK, Vaslow DF, Wang CH, Hillman RE, Farmer JE (2005) Essential Versus Complex Autism: Definition of Fundamental Prognostic Subtypes. Am J Med Genet A 135:171-180

Molloy CA, Murray DS, Kinsman A, Castillo H, Mitchell T, Hickey FJ, Patterson B (2009) Differences in the Clinical Presentation of Trisomy 21 with and without Autism. J Intellect Disabil Res 53:143-151

Morrow EM, Yoo SY, Flavell SW, et al (2008) Identifying Autism Loci and Genes by Tracing Recent Shared Ancestry. Science 321:218-223

Motazacker MM, Rost BR, Hucho T, et al (2007) A Defect in the Ionotropic Glutamate Receptor 6 Gene (GRIK2) is Associated with Autosomal Recessive Mental Retardation. Am J Hum Genet 81:792-798

Murch SH, Anthony A, Casson DH, Malik M, Berelowitz M, Dhillon AP, Thomson MA, Valentine A, Davies SE, Walker-Smith JA (2004) Retraction of an Interpretation. Lancet 363:750

Narita M, Oyabu A, Imura Y, Kamada N, Yokoyama T, Tano K, Uchida A, Narita N (2010) Nonexploratory Movement and Behavioral Alterations in a Thalidomide Or Valproic Acid-Induced Autism Model Rat. Neurosci Res 66:2-6

Noor A, Whibley A, Marshall CR, et al (2010) Disruption at the PTCHD1 Locus on Xp22.11 in Autism Spectrum Disorder and Intellectual Disability. Sci Transl Med 2:49ra68

Nordin V and Gillberg C (1996) Autism Spectrum Disorders in Children with Physical Or Mental Disability Or both. I: Clinical and Epidemiological Aspects. Dev Med Child Neurol 38:297-313

Orrico A, Lam C, Galli L, Dotti MT, Hayek G, Tong SF, Poon PM, Zappella M, Federico A, Sorrentino V (2000) MECP2 Mutation in Male Patients with Non-Specific X-Linked Mental Retardation. FEBS Lett 481:285-288

Pagnamenta AT, Khan H, Walker S, et al (2011) Rare Familial 16q21 Microdeletions Under a Linkage Peak Implicate Cadherin 8 (CDH8) in Susceptibility to Autism and Learning Disability. J Med Genet 48:48-54

Pavlowsky A, Gianfelice A, Pallotto M, et al (2010) A Postsynaptic Signaling Pathway that may Account for the Cognitive Defect due to IL1RAPL1 Mutation. Curr Biol 20:103-115

Pinto D, Pagnamenta AT, Klei L, et al (2010) Functional Impact of Global Rare Copy Number Variation in Autism Spectrum Disorders. Nature

Piton A, Michaud JL, Peng H, et al (2008) Mutations in the Calcium-Related Gene IL1RAPL1 are Associated with Autism. Hum Mol Genet 17:3965-3974

Piven J, Gayle J, Chase GA, Fink B, Landa R, Wzorek MM, Folstein SE (1990) A Family History Study of Neuropsychiatric Disorders in the Adult Siblings of Autistic Individuals. J Am Acad Child Adolesc Psychiatry 29:177-183

Rauch A, Hoyer J, Guth S, et al (2006) Diagnostic Yield of various Genetic Approaches in Patients with Unexplained Developmental Delay Or Mental Retardation. Am J Med Genet A 140:2063-2074

Ritvo ER, Spence MA, Freeman BJ, Mason-Brothers A, Mo A, Marazita ML (1985) Evidence for Autosomal Recessive Inheritance in 46 Families with Multiple Incidences of Autism. Am J Psychiatry 142:187-192

Robins DL, Fein D, Barton ML, Green JA (2001) The Modified Checklist for Autism in Toddlers: An Initial Study Investigating the Early Detection of Autism and Pervasive Developmental Disorders. J Autism Dev Disord 31:131-144

Roeleveld N, Zielhuis GA, Gabreels F (1997) The Prevalence of Mental Retardation: A Critical Review of Recent Literature. Dev Med Child Neurol 39:125-132

Rogers SJ, Wehner DE, Hagerman R (2001) The Behavioral Phenotype in Fragile X: Symptoms of Autism in very Young Children with Fragile X Syndrome, Idiopathic Autism, and Other Developmental Disorders. J Dev Behav Pediatr 22:409-417

Rolf LH, Haarmann FY, Grotemeyer KH, Kehrer H (1993) Serotonin and Amino Acid Content in Platelets of Autistic Children. Acta Psychiatr Scand 87:312-316

Ropers HH and Hamel BC (2005) X-Linked Mental Retardation. Nat Rev Genet 6:46-57

Ropers HH (2006) X-Linked Mental Retardation: Many Genes for a Complex Disorder. Curr Opin Genet Dev 16:260-269

Rosenberg RE, Law JK, Yenokyan G, McGready J, Kaufmann WE, Law PA (2009) Characteristics and Concordance of Autism Spectrum Disorders among 277 Twin Pairs. Arch Pediatr Adolesc Med 163:907-914

Rujescu D, Ingason A, Cichon S, et al (2009) Disruption of the Neurexin 1 Gene is Associated with Schizophrenia. Hum Mol Genet 18:988-996

Rumbaugh G, Adams JP, Kim JH, Huganir RL (2006) SynGAP Regulates Synaptic Strength and Mitogen-Activated Protein Kinases in Cultured Neurons. Proc Natl Acad Sci U S A 103:4344-4351

Samaco RC, Hogart A, LaSalle JM (2005) Epigenetic Overlap in Autism-Spectrum Neurodevelopmental Disorders: MECP2 Deficiency Causes Reduced Expression of UBE3A and GABRB3. Hum Mol Genet 14:483-492

Scheiffele P, Fan J, Choih J, Fetter R, Serafini T (2000) Neuroligin Expressed in Nonneuronal Cells Triggers Presynaptic Development in Contacting Axons. Cell 101:657-669

Schifter T, Hoffman JM, Hatten HP,Jr, Hanson MW, Coleman RE, DeLong GR (1994) Neuroimaging in Infantile Autism. J Child Neurol 9:155-161

Schopler E, Reichler RJ, DeVellis RF, Daly K (1980) Toward Objective Classification of Childhood Autism: Childhood Autism Rating Scale (CARS). J Autism Dev Disord 10:91-103

Schutt J, Falley K, Richter D, Kreienkamp HJ, Kindler S (2009) Fragile X Mental Retardation Protein Regulates the Levels of Scaffold Proteins and Glutamate Receptors in Postsynaptic Densities. J Biol Chem 284:25479-25487

Sharma A, Hoeffer CA, Takayasu Y, Miyawaki T, McBride SM, Klann E, Zukin RS (2010) Dysregulation of mTOR Signaling in Fragile X Syndrome. J Neurosci 30:694-702

Shoubridge C, Tarpey PS, Abidi F, et al (2010) Mutations in the Guanine Nucleotide Exchange Factor Gene IQSEC2 Cause Nonsyndromic Intellectual Disability. Nat Genet 42:486-488

Shuang M, Liu J, Jia MX, Yang JZ, Wu SP, Gong XH, Ling YS, Ruan Y, Yang XL, Zhang D (2004) Family-Based Association Study between Autism and Glutamate Receptor 6 Gene in Chinese Han Trios. Am J Med Genet B Neuropsychiatr Genet 131B:48-50

Skaar DA, Shao Y, Haines JL, et al (2005) Analysis of the RELN Gene as a Genetic Risk Factor for Autism. Mol Psychiatry 10:563-571

Smalley SL, Asarnow RF, Spence MA (1988) Autism and Genetics. A Decade of Research. Arch Gen Psychiatry 45:953-961

South M, Williams BJ, McMahon WM, Owley T, Filipek PA, Shernoff E, Corsello C, Lainhart JE, Landa R, Ozonoff S (2002) Utility of the Gilliam Autism Rating Scale in Research and Clinical Populations. J Autism Dev Disord 32:593-599

Spence SJ and Schneider MT (2009) The Role of Epilepsy and Epileptiform EEGs in Autism Spectrum Disorders. Pediatr Res 65:599-606

Steffenburg S, Gillberg C, Hellgren L, Andersson L, Gillberg IC, Jakobsson G, Bohman M (1989) A Twin Study of Autism in Denmark, Finland, Iceland, Norway and Sweden. J Child Psychol Psychiatry 30:405-416

Steiner CE, Guerreiro MM, Marques-de-Faria AP (2004) Brief Report: Acrocallosal Syndrome and Autism. J Autism Dev Disord 34:723-726

Stevens MC, Fein DA, Dunn M, Allen D, Waterhouse LH, Feinstein C, Rapin I (2000) Subgroups of Children with Autism by Cluster Analysis: A Longitudinal Examination. J Am Acad Child Adolesc Psychiatry 39:346-352

Stromland K, Nordin V, Miller M, Akerstrom B, Gillberg C (1994) Autism in Thalidomide Embryopathy: A Population Study. Dev Med Child Neurol 36:351-356

Stromme P and Diseth TH (2000) Prevalence of Psychiatric Diagnoses in Children with Mental Retardation: Data from a Population-Based Study. Dev Med Child Neurol 42:266-270

Stromme P, Mangelsdorf ME, Scheffer IE, Gecz J (2002) Infantile Spasms, Dystonia, and other X-Linked Phenotypes Caused by Mutations in Aristaless Related Homeobox Gene, ARX. Brain Dev 24:266-268

Sultana R, Yu CE, Yu J, et al (2002) Identification of a novel gene on chromosome 7q11.2 interrupted by a translocation breakpoint in a pair of autistic twins. Genomics 80:129-134.

Symons FJ, Sperry LA, Dropik PL, Bodfish JW (2005) The Early Development of Stereotypy and Self-Injury: A Review of Research Methods. J Intellect Disabil Res 49:144-158

Tahiliani M, Mei P, Fang R, Leonor T, Rutenberg M, Shimizu F, Li J, Rao A, Shi Y (2007) The Histone H3K4 Demethylase SMCX Links REST Target Genes to X-Linked Mental Retardation. Nature 447:601-605

Tarpey PS, Raymond FL, Nguyen LS, et al (2007) Mutations in UPF3B, a Member of the Nonsense-Mediated mRNA Decay Complex, Cause Syndromic and Nonsyndromic Mental Retardation. Nat Genet 39:1127-1133

Toth K, Dawson G, Meltzoff AN, Greenson J, Fein D (2007) Early Social, Imitation, Play, and Language Abilities of Young Non-Autistic Siblings of Children with Autism. J Autism Dev Disord 37:145-157

Tuchman R and Rapin I (2002) Epilepsy in Autism. Lancet Neurol 1:352-358

Turner G, Partington M, Kerr B, Mangelsdorf M, Gecz J (2002) Variable Expression of Mental Retardation, Autism, Seizures, and Dystonic Hand Movements in Two Families with an Identical ARX Gene Mutation. Am J Med Genet 112:405-411

Tzschach A, Lenzner S, Moser B, et al (2006) Novel JARID1C/SMCX Mutations in Patients with X-Linked Mental Retardation. Hum Mutat 27:389

Viding E and Blakemore SJ (2007) Endophenotype Approach to Developmental Psychopathology: Implications for Autism Research. Behav Genet 37:51-60

Vincent JB, Horike SI, Choufani S, Paterson AD, Roberts W, Szatmari P, Weksberg R, Fernandez B, Scherer SW (2006) An Inversion Inv(4)(p12-p15.3) in Autistic Siblings Implicates the 4p GABA Receptor Gene Cluster. J Med Genet 43:429-434

Volkmar FR, Szatmari P, Sparrow SS (1993) Sex Differences in Pervasive Developmental Disorders. J Autism Dev Disord 23:579-591

Wakefield AJ, Murch SH, Anthony A, et al (1998) Ileal-Lymphoid-Nodular Hyperplasia, Non-Specific Colitis, and Pervasive Developmental Disorder in Children. Lancet 351:637-641

Wang K, Zhang H, Ma D, et al (2009) Common Genetic Variants on 5p14.1 Associate with Autism Spectrum Disorders. Nature 459:528-533

Wassink TH, Piven J, Patil SR (2001) Chromosomal Abnormalities in a Clinic Sample of Individuals with Autistic Disorder. Psychiatr Genet 11:57-63

Wing L and Gould J (1979) Severe Impairments of Social Interaction and Associated Abnormalities in Children: Epidemiology and Classification. J Autism Dev Disord 9:11-29

Witwer AN and Lecavalier L (2007) Autism Screening Tools: An Evaluation of the Social Communication Questionnaire and the Developmental Behaviour Checklist-Autism Screening Algorithm. J Intellect Dev Disabil 32:179-187

Yeargin-Allsopp M, Rice C, Karapurkar T, Doernberg N, Boyle C, Murphy C (2003) Prevalence of Autism in a US Metropolitan Area. JAMA 289:49-55

Zeegers M, Van Der Grond J, Durston S, Nievelstein RJ, Witkamp T, Van Daalen E, Buitelaar J, Engeland HV (2006) Radiological Findings in Autistic and Developmentally Delayed Children. Brain Dev 28:495-499

Zemni R, Bienvenu T, Vinet MC, et al (2000) A New Gene Involved in X-Linked Mental Retardation Identified by Analysis of an X;2 Balanced Translocation. Nat Genet 24:167-170

Zweier C, de Jong EK, Zweier M, et al (2009) CNTNAP2 and NRXN1 are Mutated in Autosomal-Recessive Pitt-Hopkins-Like Mental Retardation and Determine the Level of a Common Synaptic Protein in Drosophila. Am J Hum Genet 85:655-666

Part 4

Treatment and Genetic Counseling

9

Microgenetic Approach to Therapy of Girls with ASD

Katarzyna Markiewicz[1] and Bożydar L.J. Kaczmarek[1, 2]
[1]*University of Maria Curie Sklodowska*
[2]*Szczecin University*
Poland

1. Introduction

The term Autism Spectrum Disorders (ASD) came into use in 1988 to stress the fact that symptoms of autism may appear with various intensity – from very mild to very severe. The matter gets further complicated due to the fact that the symptoms occur in different combinations. Autistic deficits include not only "autistic triad", that is disorders of speech, behaviour and social interactions but disturb also cognitive, and motor abilities as well as emotional functioning. Despite a long-term interdisciplinary studies and detailed descriptions of autistic symptoms diagnosis of ASD still causes a number of diagnostic difficulties. The diagnostic procedure has been made easier thanks to the introduction of two systems of classification - International Statistical Classification of Diseases and Related Health Problems (ICD-10)[1] and DSM-IV-TR[2] - but it still remains a long and arduous process. It happens quite often that the final diagnosis is made in the course of therapy.

Assessment of children with ASD must takes into account their spontaneous and reactive behaviours as well as information gained during the interview with their parents. As mentioned above, dysfunctions occurring in autistic spectrum vary to a considerable degree in their intensity. Moreover, they are of a dynamic character and are apt to undergo changes as a result of a course of general development, individual experience, social conditions, undertaken therapy, and efficacy of stimulation. All these may hamper the process of evaluation, especially at the early stages of a child development. Hence, autistic children are often made a diagnosis of mental disability, behaviour disorders, hearing problems as well as strange and eccentric behaviours. To complicate the matter further the above disorders frequently coexist with autism. It is imperative, therefore, to be able to discriminate autism and other developmental disorders since the early diagnosis provides basis for creating effective therapeutic and educational programs.

2. Frequency of ASD occurrence

Recent epidemiological studies show a dramatic increase of the number of persons with autistic symptoms. Thus, the frequency of occurrence of ASD was rated at a level of 4 to 10,000

[1] Proposed by World Heath Organization – WHO, 10th revision, 1992

[2] Diagnostic and Statistical Manual of Mental Disorders developed by American Psychiatric Association - APA, 2000; a new version V is to be published in May 2013

cases in the years of 1980-90, while in the last decade the number of autistic persons is considered to reach the level of 2 to 1000 persons, and 2 to 1000 persons in the case of Asperger syndrome. At the same time, the population of persons with ASD is rated at the level of 1-2 to 100 cases in the U.S.A., the country in which the most advanced epidemiological studies are carried out. In addition, statistical data show that the increase of a frequency rate of ASD is 10 - 17% a year (Newschaffer et. al., 2007). It may be due to a real increase of the occurrence of autism but might as well reflect refinement of assessment techniques as well as a wider scope of diagnostic criteria, such as lowering down the age of assessment, broadening the scope of diagnostic criteria as well as development of diagnostic tools and techniques. In addition, the awareness of the parents of disabled children has also changed lately.

An analysis of statistical data shows also that ASD disorders are more frequent in boys than in girls. It is estimated as 4,3 to 1, while in the case of Asperger syndrome the occurrence of the disorder in boys rises to 8 in comparison to 1 girl (ICD-10, 1997). There are a number of theories trying to explain such a state of affairs. One of frequently cited is the theory of neurotoxic testosterone. It presumes that due to the unequal development of brain hemispheres in the prenatal stage a very high level of testosterone brings about disturbances in the development of the left hemisphere, which results in a very high occurrence of autistic disorders in boys (Geschwind, Galaburda, 1985a, 1985b, 1985c).

The idea was further elaborated by Baron-Cohen and collaborators (2005). They conducted a longitudinal study, which aimed at revealing a possibility of the influence of so called foetal testosterone (FT) upon the development of a child in the prenatal stage of life. The study included pregnant women in whom level of testosterone was measured in amniotic fluid obtained via amniocentesis in order to evaluate the influence of difference in a level of FT upon a subsequent development of the child. The results described also in the following paper (Auyeung et al., 2009) revealed a negative correlation between FT and the development of language and social skills, and positive correlation with specific traits of autism, such as excessive concentration on details, stereotyped movements, and perseverations. The studies on hiper-mascunalization in ASD are carried on, since the role of the influence of testosterone upon the foetal development needs further verification. Yet, an analysis of androgens influence on the appearance of autistic disorders seems to bring promising results (Chakrabarti et al., 2009).

The above presented data raises a question whether a clinical picture of autistic symptoms in girls differs from that observed in boys, and what - if any - is a specificity of autism occurring in girls. Kopp (2010) noted that autistic symptoms in girls are often neglected by professionals despite the reports of anxious parents. It creates the need of presenting a more detailed analysis of autistic disorders observed in girls. All the more that the most available descriptions concentrate upon the characteristics of autism based upon an analysis of boys are of a rather general nature. Therefore, we decided to perform an analysis of autistic symptoms in autistic girls from the perspective of microgenetic and microdevelopment approaches. We were motivated by the fact that those theories make possible the evaluation of developmental potential of autistic persons. In order to make the clinical picture of autism in girls as complete as possible the description of individual cases will be presented in some detail.

3. Autism spectrum disorder in girls

As mentioned above, most studies published so far give general information, which is mostly based upon an analysis of autistic boys. As a rule the sex of examined subjects is not taken into consideration, especially as ASD is prevailing in boys. Lately, specificity of ASD occurring in girls has been noted, yet most authors still concentrate upon symptoms characteristic of autism in general, such as tendency to routine behaviours, lack of interest in fantasy games, difficulties in social interactions and/or distractibility leading to learning problems (Knopp, 2010). On the other hand, McLennan and collaborators (1993) report that Loveland observed lowering of general IQ in boys (N = 700), while it was not stated in girls (N = 300). At the same time, it was noted that no use is made of the developmental potential of autistic girls, which may result in lower social and communication capacities and greater difficulties in establishing relations with peers in girls (Lord & Schopler, 1985). In addition, girls score worse than boys both in verbal abilities and visual-spatial tests. According to Nichols (2009) it reflects higher level of expectations concerning social and communication skills in girls, and hence more negative evaluation of observed deviations. Inappropriate behaviours of girls are often interpreted as a way to make others to pay attention to them, while in the case of boys such behaviours are believed to reflect their attempts to get a desired object.

The differences between sexes are also believed to be a result of differences in developmental trajectories. Some authors believe that boys exhibit more difficulties at the early stage of their life, while the difficulties in girls increase in the period of adolescence (Nichols, 2009; Nichols et al., 2008). Symptoms of brain damage, however, are more frequent in girls than in boys, which finds its confirmation in EEG records, which show more irregularities in girls. On the other hand, the autistic girls are better in performing games that require using rules and also show weaker tendency to stereotyped movements than boys (Lord et al., 1982; Nichols, 2009).

Lord and collaborators (1985) point out that longitudinal studies revealed that girls with ASD did not establish any friendly relationships during a period of ten years, while several boys did accomplish it. According to these authors it may be due to the fact that girls tend to be more short-tempered and tearful (see also McLennan et al., 1993). Those discrepancies in behavioural and neurological functioning tend to disappear if the autistic girls are offered appropriate stimulants for their development. Yet, the autistic symptoms often appear again as the years go.

Kopp and colleagues (2010) compared the quality of life of 100 girls with ASD and ADHD aged from 3 to 18 years, and it made them believe that those two types of disorders tend to co-occur, since ADHD was stated in 95% of autistic girls. At the same time, a higher level of fear, sleeping problems as well as a higher risk of depression was noted in both ASD and ADHD. Moreover, comparative studies of girls with ASD and ADHD revealed a regression in development in comparison to healthy subjects of the same age. The dysfunctions were observed both in psychological, motor, and social abilities so they included all aspects of behaviour. It needs to be stressed that a positive influence of environmental factors, especially of appropriate education and family conditions, proved to stimulate the development of autistic girls. Hambrook and collaborators (2008) observed that anorexic girls exhibit lack of empathy and of an ability to systemize as well as other autistic traits. It is emphasized that a distorted pattern of information processing characteristic of anorexia shows a significant similarity to the autistic spectrum. Those difficulties may take various forms such as a lack of cognitive flexibility or stereotyped behaviours. A good example of rigid patterns of response noted both in ASD and anorexia provides an inability to shift a plan of action.

4. Microgenetic approach form developmental point of view

A traditional approach to the study of developmental processes concentrates either on a long period of time (longitudinal studies) or upon groups including as big numbers as possible (cross-sectional studies). In recent years many scholars emphasize the usefulness of research taking into account the scores gained by the same group of children evaluated in short time intervals. Such an approach has derived also from the microgenetic theory. In other words, it enables an observer to monitor the specific moments of transformations in thought and behaviour in contradistinction to classic longitudinal studies, which provide only a general pattern of a change of behaviour in examined subject (Levelly et al., 2005).

Flynn and co-workers (2006) enumerate three aspects pointing to the usefulness of a microgenetic approach:

1. It makes possible delineation of a whole range of a mechanism underlying a process of changes.
2. Observations are conducted while the factor causing a change is at work and not only before and after it took place.
3. It is possible to control a moment of passing from applying a stimulus and initiating a change.

At the same time, microgenetic approach enables getting answers to the following questions:

1. Is the instruction understood by the child?
2. Does the child use innovative strategies in solving a given problem?
3. Is the child able to discover a new more effective strategy in the course of action?
4. What is the efficiency of actions undertaken by the child while looking for a proper solution?
5. What amount of time does a child need to solve a particular problem?
6. Is the child able to generalize an acquired strategy to solve other similar problems?
7. In what way was a new experience acquired?

Microgenetic approach aims at an analysis of changes occurring during solving a given problem that takes into account five dimensions of cognitive growth. They include path, rate, breadth, source, and variability (Siegler & Svetina, 2002; Calais, 2008)).

- The **path** of change involves the sequence of problem solving attempts performed by children in their way to gaining required competence. It also shows if the change is qualitative or quantitative.
- The **rate** of change concerns the time or amount of experience the children needed to start using a new strategy in a consistent way. It also includes an analysis of the nature of a change - whether it occurred gradually or suddenly.
- The **breadth** of change reflects children's ability widely to generalize a new approach to other problems and contexts.
- The **source** of change takes into account factors that evoked observed changes.
- The **variability** of change enables evaluation of individual traits of a child in acquiring other dimensions of change. In other words, it enables creating a characteristics of an individual child as well as comparing a pattern of change across individuals (Siegler & Svetina, 2002; Flynn et al. 2006; Calais, 2008)).

It is due to the fact that concentration upon the process of change as it is occurring reveals what mechanisms underpin it. It thus makes possible identifying both detrimental changes (areas of dysfunctions) as well as positive ones reflecting a developmental potential of a given child.

5. Microgenesis from neuropsychological point of view

A neuropsychological perspective of the microgenetic theory points to the fact that each action starts at lower levels of the brain and unfolds to the higher more specialized levels. It enables a fresh look upon the nature of symptoms observed in brain dysfunctions. As pointed out by Brown & Pachalska (2003) "the lesion displays phases in a transitional sequence from depth to surface" (p. 4). Two important notions are introduced here: parcelation and heterochrony. Parcelation means the elimination of cells and connections, which occur in over-abundance at birth, in order to achieve specificity. It is connected both with maturation and cognitive experience. Hence, sensory deprivation results in a diffuse and redundant connectivity and a loss of the ability to discriminate among perceived stimuli.

In the case of function the same role as elimination is played by inhibition. Brown and Pachalska state: "Inhibition occurs in the development of action, in newborns, which goes from global movement of the hand or face (the cherubic face of the infant) to one that is more finely individuated" (2003, p. 6). In other words, the basic pattern of each system are elimination, inhibition and specification, which means sculpting away constraints at successive phases of cognitive activity. The authors point out that in pathology re-generalization through disinhibition leads to a number of deviant behaviours.

Another important notion of microgenesis is heterochrony. It assumes that the fact that different brain systems develop at different rates can result not only in adaptations to the environment but also in malfunctions and aberrations. It is connected with the phenomenon of neoteny, which means selective prolongation of an immature phase of development. It makes possible refinement of structures and functions allowing mastering higher cognitive processes of which verbal communication is a good example.

As pointed out by Brown (1998, 2001) microgenesis assumes that phyletic and ontogenetic growth patterns are retraced in microgeny but the processes are collapsed here over a second or in a fraction of a second. Moreover, both mental and motor processes have a hierarchical structure as the later levels unfold out of earlier ones. It is, therefore, possible to analyze the changes in children's behaviour when they attempt to solve a given problem passing from one level (or stage) to another. And that is of much help in making a course of therapy as effective as possible.

6. Procedure

The aim of our study was to delineate the cognitive abilities of autistic girls in the context of the microgenetic approach. The following question was asked:

What are the characteristic traits of the cognitive development in girls?

In order to find an answer to the above formulated question a detailed description of changing competence of three girls with ASD will be presented. All the three girls represented similar level of autistic features and of cognitive abilities. All of them were able to communicate verbally and all of them were diagnosed in accordance with ICD-10 (International Statistical Classification of Diseases and Related Health Problems) criteria.

The control variables were: depth of autistic deficits, and the level of social, and communicative competence as well as interest in playing and reacting to applied incentives. The elapse of time between individual examinations was also controlled. The boys development was presented in other works (Markiewicz, 2008, 2009; Markiewicz & Pachalska, 2007; Markiewicz & Grochmal, 2008; Markiewicz & Mc Queen, 2008;; Markiewicz et al., 2009), therefore their results will be used as a background for the description of girls.

6.1 Case one

Ola B. was first diagnosed at the age of three with the suggestion of middle stage mental disability. No neurological dysfunctions were noted. But her environmental conditions were very bad since her mother died when the girl was 3 months old and her father was an alcoholic. His parental rights were juristically limited when the girl was 2 years old, and she was at the custody of her grandmother (the mother of mother). The first verification of the initial diagnosis was after the girl was 4 years old at our clinic this time. The assessment performed by our team indicated dysfunctions in all spheres of developmental development (F84) suggesting Asperger syndrome (F84.5). The results of psychological examination are presented below.

6.1.1 Results from ICD-10

The evaluation of behaviours performed with ICD-10 revealed:

1. Qualitative impairments in social interactions, which was manifested by:
    a. Difficulties in accepting new situations (e.g. signs of frustration if a sequence of known activities was changed), limited social activity (she did not undertake interactions on her own but undertook simple forms of activity if initiated by an adult)
    b. Weak adaptability to surrounding stimuli, mainly social ones (e.g. she was entirely indifferent to the new persons in her environment)
    c. Emotional distance, lack of spontaneous expression of feelings with the use of speech, gestures, and facial expressions.
2. Communication disabilities revealing in:
    a. Lack of initiating verbal contacts, and limited readiness to communicate
    b. Weak reactions to questions and commands with preserved understanding (she performed simple verbal commands such as 'come here')
    c. Weak reactions to visual and/or auditory stimuli, and limited reactivity to non-linguistic messages
    d. Speech limited to simple sentences that often were constructed against grammatical rules
    e. Limited inventory of instrumental gestures and emblems (which commonly are used in place of speech).
3. Behaviour disorders that revealed in:
    a. Many stereotyped reactions (turning round on tiptoes, swinging, beating a floor with an object, and non-functional use of objects)
    b. Lack of correct reaction to stimuli as well as inability to differentiate reactions in response to various character of stimuli, odd treatment of objects, making dices and blocks move.

c. Lack of emotional reactions to new toys characteristic of young children (lack of curiosity and pleasure connected with receiving a toy, interest in new objects)
d. Lack of initiative to start playing.

6.1.2 Results from PEP-R (psychoeducational profile – revised)

The scores of Ola on the developmental scale are typical of 30-34 months of age (she was 48 months old at that time). After taking into account emerging scores they rise to 43-47 months of age, which makes it closer to the age of 3years and 9 months. The highest scores the girl gained in gross motility. At a similar level, yet significantly below her age, were skills of imitation, perception, fine motor, and an ability to come into relationships (relating and effect), sensory responses, and eye-hand coordination. Among them most promising proved to be sensory responses, eye-hand coordination, and cognitive processes with the exclusion of language. Therefore, those three spheres were regarded as the zone of proximal development.

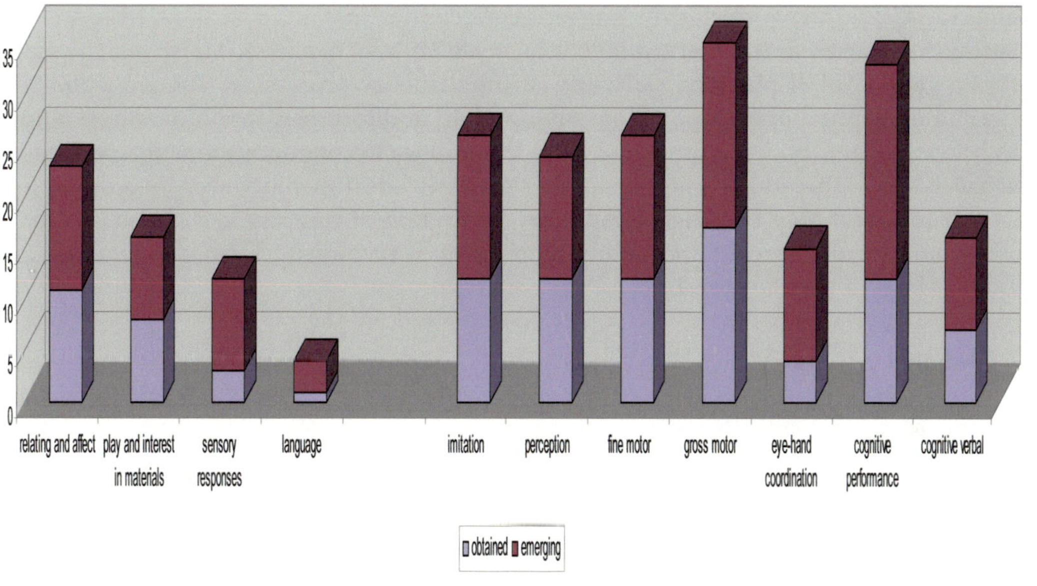

Fig. 1. Obtained and emerging scores of Ola on PEP-R (Psychoeducational Profile - Revised)

Consequent examinations confirmed occurrence of the so called autistic triad, that is trouble getting along with others in a social circle, especially when working together, sharing feelings and thoughts, and making friends; impairments of communication; impairments of flexible imaginative functions of which restricted and repetitive behaviours and interests as well as difficulties in coping with changes are most characteristic. A significant variable, which might have influenced the development and functioning of the girl, was her traumatic babyhood experience. Yet, the character of developmental changes allows the conclusion that it had been rather a secondary factor, though the experiences of her early childhood might have stimulated the appearance of ASD dysfunctions.

6.1.3 Results from WISC-R (Wechsler intelligence scale for children – revised)

The next examination connected with a need to assess girl's school readiness at the age of 6;9 revealed a considerable range of scores. They ranged form moderate intellectual disability to average scores. Her general IQ was in the range of <48 - 76>; the scores of verbal IQ were within <40 - 58>; while those of performance scale rose to <73 - 98> ($p = 0.05$). The highest scores in the verbal scale the girl got in arithmetic - they were within the average. Other scores were at the range of low and very low. In the case of the performance scale it were the block design, object assembly, and coding, which were scored at the average level. Close to average were also her scores on picture completion and picture assembly.

It is also worth pointing out that the girl was able to comprehend simple commands while fulfilling tasks connected with examination, and she also came into verbal contact with the examiner. A three-way analysis revealed that only perceptual organization scored at average level (7.75), resistance to distracters scored below average (5), while scores of comprehension were very low (1).

Thus, the scores gained by the examined girl revealed average abilities in visual analysis and synthesis as well as good perception of abstract stimuli, and good long-term visual memory. At the same time they suggest correct imagination, visual orientation, and visualization as well as a preserved ability to create abstract concepts. Moreover, the scores reflect a good level of planning skills and of simultaneous processing. The girl's ability to concentrate attention upon a particular task, and an auditory perception of simple stimuli was below average. But most impaired were the abilities to communicate while performing particular tasks. This might suggest that her auditory short-memory was disordered (She was able to repeat only two digits). Auditory perception of complex verbal stimuli, verbal expression, verbal memory, and ability to remember previously learned utterances also proved to be on a very low level.

6.2 Case two

Monika J. was first diagnosed before she was 3 years old. An intellectual retardation of a moderate level (F71) with "some traits of autistic behaviours" was stated at the other clinics. The verification of this diagnosis performed by our team two years later pointed to persuasive developmental disorders (F84), autism in particular (F84.0).

6.2.1 Results from ICD-10

The evaluation of her behaviour in accordance with the criteria of ICD-10 showed:

1. Qualitative impairment of social interactions:
    a. Inadequate evaluation of social and emotional signals, incorrect reactions to the emotional states of others, weak modulation of behaviour in response to a given social context.
    b. Low level of social skills, weak integration of social, emotional, and communicative behaviours
    c. Disturbances in reciprocal social interactions.
2. Restricted, stereotyped , and repetitive interests and activities:
    a. Routine and inflexibility in everyday behaviours, and making others stick to such actions
    b. Stereotyped movements

c. Attachment to the computer appeared with age, including the interest in its construction.

3. Disorders of communication:
   a. Low level of social use of language, weak synchronization and lack of reciprocation in dialogue
   b. Weak changeability of verbal expression; limited ability emotionally to modulate utterances, and weak reactions to questions and instructions as well as to nonverbal clues.
   c. Difficulties in differentiation of rhythm and accent to modulate communication
   d. Limited ability to use facial expressions and gestures in communication.

6.2.2 Results from PEP-R

The scores of PEP-R show that the actual level of the girl's abilities is typical of the age range 22 - 29 month both in the developmental and behavioural scale. It means that they are significantly below her chronological age as she was 52 months old at the moment of examination. She scored best in gross motor, fine motor, cognitive performance, and perception, while the scores of imitation, language, cognitive verbal, relating and affect as well as play and interest in materials were significantly below average. At the same time, her emerging scores indicated abilities typical of the age range 46 to 51 months so they were close to her chronological age. It was observed in accomplishing such tasks as cognitive performance, imitation, relating and affect as well as play and interest in materials, providing for her zone of proximal development. Scores gained by Monika during an examination with the use of PEP-R are presented in figure 2.

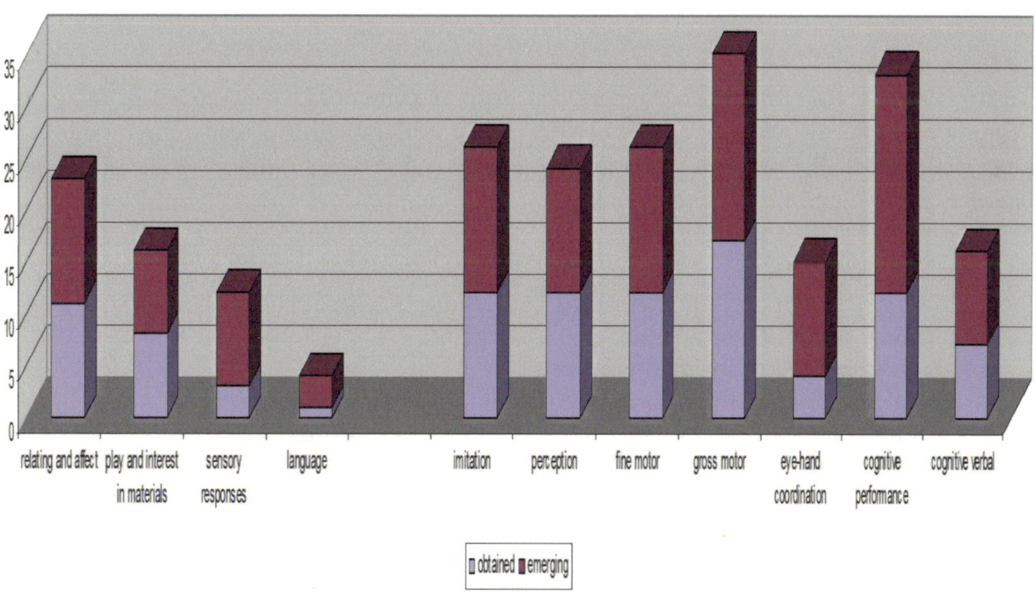

Fig. 2. Obtained and emerging scores of Monika on PEP-R (Psychoeducational Profile - Revised)

6.2.3 Results from WISC-R

The examination administered at the age of 6 years 11 months for the needs of evaluating her school readiness revealed a wide range of scores form moderate mental retardation to an average level. Her general IQ was within the range of <40 - 68> , the verbal IQ was within <34 – 53>, and the performance scores were within the range of <65 – 85>. She accomplished the following subtests of Wechsler Scale on the average level: object assembly, picture arrangement, and block design, while on similarities, vocabulary, and arithmetic subtest she scored very low. It should be pointed out here that despite her difficulties with defining terms the girl was quite good in communicating with others. She was able to understand simple commands and to carry on conversation. She was also surprisingly good at reading messages displayed on a computer screen. She was also able to solve simple puzzles and to put together mixed verses of a poem into a meaningful whole. Moreover, she arranged individual words into sentences, and knew the meaning of road signs. Her arithmetic abilities were also quite good as she recognized and wrote down digits starting from 0 to 100, and she was able to compare digits within the range of ten with the use of signs: =, <, >.
Three-way analysis revealed that her perceptual organization was at an average level, while the other factors, such as reasoning (1.5) and resistance to distracters (1) were on a very low level. Her scores suggest average level of visual analysis and synthesis, and of visual perception as well as a correct long-term memory. They also indicate good visual orientation and visualization as well as preserved ability to create abstract concepts. Her abilities to plan and simultaneous processing were also preserved.
However, her abilities to concentrate upon a particular task and to interact verbally while performing it as well as her auditory perception of simple stimuli were limited. Very severely disturbed was the perception of complex stimuli, verbal expression, durability of verbal memory, and an ability to remember previously learned expressions.

6.3 Case three

Gabriel R. was first diagnosed when she was 3 years old. It suggested delay and disharmony of her development. The verification of initial diagnosis was done by our team when she reached the age of 4 years and 4 months. It indicated persuasive developmental disorders (F84) suggesting autism (F84.0).

6.3.1 Results from ICD-10

The evaluation of her behaviour performed in accordance with ICD-10 revealed:

1. Impairment of social interactions:
   a. Lack of reciprocal social interactions
   b. Weak modulation of behaviour in response to a given social context
   c. Lack of interest in social aspects of play and/or performing tasks
   d. Lack of spontaneous interactions with the closest.
2. Restricted repertoire of interests and activities:
   a. Sniffing at the surrounding, liking of sharp odours
   b. Stereotyped behaviours, turning round on tiptoes, very quick walking on tiptoes
   c. Obsessive interest in maps and diagrams of technical devices.
3. Decreased communicative competence

a. Lack of ability to use the language in various social interactions
b. Inadequate reactions to verbal cues

c. Monotonous non-modulated utterances with agrammatisms
d. Difficulties with understanding complex utterances
e. Difficulties with synchronization of linguistic and non-linguistic aspects of communication.

6.3.2 Results from PEP-R

The girl's scores are at the age range of 37-42 month of life. They are, therefore, below her chronological age. If we take into account emerging scores the age range rises to 50-54 months of life, which means that her potential abilities are close to the chronological age. A detailed analysis reveals that the girl scored best in gross motor, fine motor, cognitive performance, eye-hand coordination, and perception in developmental subscales, while in the case of behavioural scale she scored best on play and interest in materials, and sensory responses. She obtained significantly low scores on language, and relating and affect (behavioural scale) as well as on cognitive verbal, and imitation (developmental scale). The zone of proximal development included relating and affect, language, imitation, and cognitive verbal subscales. The scores of PEP-R test administered at the age of 4 years and 4 months are presented in figure 3.

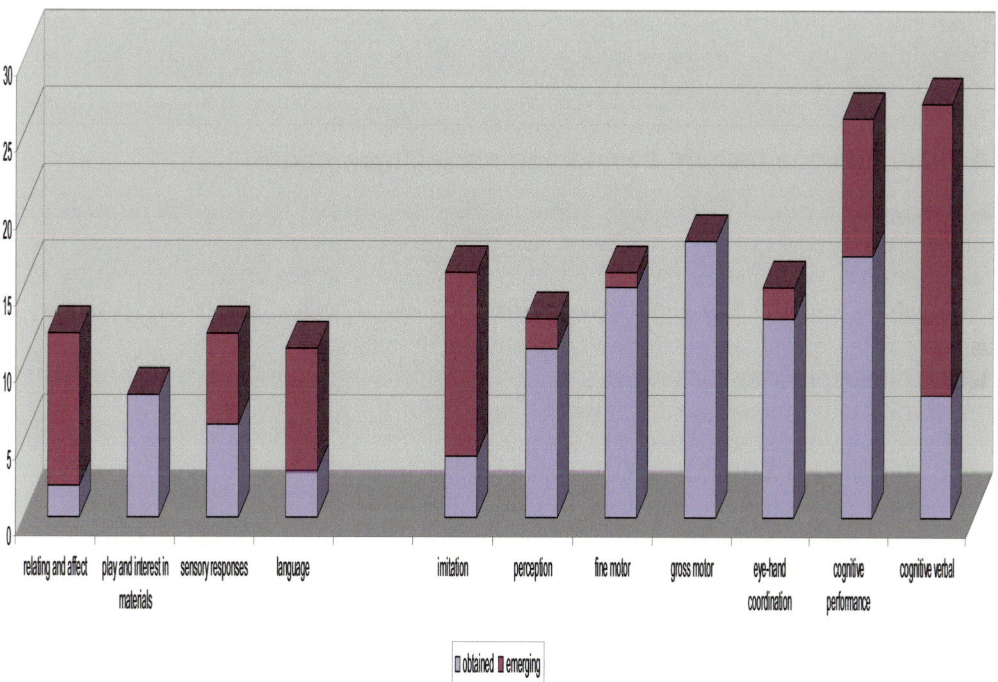

Fig. 3. Obtained and emerging scores of Gabriel on PEP-R (Psychoeducational Profile - Revised)

6.3.3 Results from WISC-R

Next stages of diagnostic-therapeutic procedure confirmed the existence of an autistic triad. At the same time, an examination of school readiness at the age of 6 years 11 months indicated the distribution of scores within the range from moderate mental retardation to below average. The scores were within <56 - 74> for the verbal scale, <63 - 83> for the performance scale, and <52 - 80> in the case of the full scale IQ (confidence level $p = 0.05$). The highest verbal IQ scores (ranging within average results) the girl gained on arithmetic and information. The remaining scores were very low. As far as the performance IQ is concerned she scored on average level on picture completion, block design, and object assembly subtests. On the remaining subtests her scores were below average (picture arrangement) or very low (coding). She also exhibited difficulty in defining concepts (on vocabulary subtest) in a way similar to the two above described autistic girls. She was, however, quite good at comprehending simple commands and instructions.

To give the reader some insight into the nature of the girl's difficulties we give some samples of her responses appearing in the vocabulary subtest: 'umbrella' – like rain; 'clock' – bim-bam; 'alphabet' –a,b,c…(she made an attempt to enumerate the letters of the alphabet); 'nail' – it is cut or not cut; 'bicycle' – she lied down on the floor and showed how to ride a bicycle; 'knife' – she performed the movements of cutting.

Three-way analysis revealed a correct level of development of perceptual organization (7.5) and of resistance to distracters (8.7). Yet, her score on verbal reasoning was very low (2.7). It may reflect her low level of linguistic and communicative competences. On the other hand, the above scores point to good visual processing, and preserved ability to concentrate on a given task. Her abilities to create abstract concepts as well as planning and simultaneous processing were also quite good.

7. An example of a task on developing social interactions

The microgenetic model makes it possible to directly observe the acquisition of cognitive abilities in the process of training. Below we give an example of a task that made possible the evaluation to what extent the child is able to comprehend the visual perspective of both her and of the parent. The girls were to demonstrate to the parent a toy that they had constructed by themselves. The following elements we assessed:

1. The child approaches the parent
2. The child tells the parent: "Look what I have done"
3. The child demonstrates the toy to the parent
4. The child moves her eyes from the toy to the parent
5. The child moves her eyes from the parent to the toy
6. The child demands a commendable: "Do you like it"

During the experiment the examiner and the child were seated at a table, and the parent was seated at another table at the opposite corner of the room. The child was able to see the whole room, while the parent was seated diagonally at the other side. The examiner presented the child a puppet consisting of blocks strung on a stick with a stand. The blocks could be arranged in various configurations so that the shape of the puppet was changed. The girl constructed the puppet, and then the examiner said:

1. Show me what you have done
2. Look at the puppet
3. Do I look at the puppet?
4. Do you like the puppet?
5. Do I like the puppet?

After a short period of joint play, the examiner handed the child the puppet and said: "Show the puppet to your mum/daddy". The command was accompanied by pointing to the parent with a hand of the examiner. The parents had been instructed to: (1) put their hands over their eyes, (2) sit with their backs to the child. Hence, in order to demonstrate the puppet the child had to approach the parent and make her look at the puppet, for example by pulling her hand and saying "Look what I have done". The task was evaluated in the following way:

0 points - no demonstrating the puppet to the parent

1 point - the child approaches the parent but keeps the puppet and looks at it without demonstrating it to the parent

2 points - the child demonstrates the puppet to the parent with an appropriate statement.

Altogether six tasks were performed during one session, and the possible maximum score was 12 points. The same pattern of six tasks was then repeated during three consecutive sessions presented at two weeks intervals, and the analysis took into account a total of all scores gained by the examined girls during all sessions.

8. Results

The scores of the girls described in the present chapter are presented in table 1. They show that all of them mastered a particular schema of action. During the first session only one girl approached her parent in the second task but she did not encourage her/him to take interest in the toy. The remaining two girls reached those criteria only in the fifth task. The qualitative analysis, however, will make it possible to delineate the strategy of forming reciprocal social interactions in a task schema by girls with ASD. After the first session the examiner demonstrated the flow diagram. The demonstration was to show the child what she was expected to do. During the second session the girls were shown pictures, in which consecutive stages of actions to be done were presented. The pictures were shown before each task.

		Task 1	Task 2	Task 3	Task 4	Task 5	Task 6	Total
Session I	Ola	0	0	0	0	1	1	2
	Monika	0	0	1	1	1	1	4
	Gabriel	0	0	0	0	1	1	2
Session II	Ola	1	1	1	1	1	1	6
	Monika	1	1	2	2	2	2	10
	Gabriel	1	1	2	2	2	2	10
Session III	Ola	1	2	2	2	2	2	11
	Monika	2	2	2	2	2	2	12
	Gabriel	2	2	2	2	2	2	12

Table. 1. Scores obtained by the examined girls in consecutive sessions.

Below we present the scores of boys in accomplishing the above described task (see table 2). Their results from Wechsler Scale were within average and lower than average range. It means that the intellectual level of the boys corresponded with intellectual abilities of the above described girls. It must be stressed, however, that all boys exhibited aggressive behaviours, such as screams, squeaks, throwing objects, tearing paper, psychomotor agitation, disobedience, and ignoring commands and prohibitions.

		Task 1	Task 2	Task 3	Task 4	Task 5	Task 6	Total
Session I	Paul	0	0	0	1	0	1	2
	Jarek	0	0	0	0	0	0	0
	Martin	0	0	0	0	1	0	1
Session II	Paul	0	0	1	1	1	2	5
	Jarek	0	1	0	1	1	1	4
	Martin	1	1	1	1	1	1	6
Session III	Paul	1	1	1	2	2	2	9
	Jarek	1	1	1	1	1	2	7
	Martin	1	1	1	2	1	2	8

Table 2. Scores obtained by boys in consecutive sessions.

Comparison of scores obtained by girls an boys shows that :

1. Results of the first sessions were similar to those of girls.
2. Difference appeared in the third session since the performance of the boys was much worse than of girls. They did not support the demonstration of the puppet with an appropriate statement.
3. Negativistic and/or aggressive behaviours were much more frequent in boys. They exhibited strong tendency to scream, to squeak, and were apt to break accomplishing the task.

8.1 Microgenetic analysis

The developmental microgenetic model (Siegler, Svetina, 2002; Calais, 2008) makes it possible to take into consideration five mentioned earlier dimensions reflecting the manner of solving a given problem. They are the path, rate, breadth, source, and variability of changes that occurred in consequence of accomplishing a particular task.

1. An analysis of the **path** revealed that the changes appearing at consecutive stages of performing a task were mainly of quantitative nature: starting from the lack of demonstrating the puppet to the parent in the initial tasks of sessions, then approaching the parent without demonstrating it, and finally demonstration with a verbal message. It allows the conclusion that the change of competence took place there resulting in an ability to come into social interaction with parents, and to understand their point of view. As can be noted in table 1, two of the girls were able to improve their action already in the third task. The girls approached their parents, and made them look at the puppet. It is worth to remind here that PEP-R examination indicated high emerging scores in imitation subscale in all the three girls. Therefore, the above tasks made use of a significant zone of proximal development.
2. An analysis of the **rate** of change showed that a difference between the primary and consequent strategy of solving a problem came into being only in the second session of

our study two weeks after the first session. However, it was not possible to state if the new strategy appeared gradually or all of a sudden. Stating it would require a more frequent, everyday, repetition of experimental sessions.

3. As far as the **breadth** of change is concerned it seems that the girls were not able to modify once learned schema of behaviour. They were not acting in a spontaneous way, and did not express any emotional commitment. Perhaps the use of other experimental strategy would enable a deeper insight into the problems encountered by the examined girls.
4. It was also stated that a carefully designed experimental procedure enables forming desired, and at the same time quite complex, actions in autistic girls. In this case the **source** of change was demonstration of expected behaviours accompanied by verbal as well as nonverbal clues.
5. It should be stressed, however, that the girls acted in an automatic way. They did not express any satisfaction with their work or a will to boast about their success. At the same time, they tended to stick to the once learned schema of action. It allows the conclusion that there was no **variability** in their behaviour as no personal traits could be noted in their manner of accomplishing the tasks as well.

Yet, it was experimentally stated that an appropriate therapeutic approach makes it possible developing schemas of social interactions. It results in developing independence and a sense of agency so important for creating a sense of identity in a given child. It is also worth to remind that all the three girls had a high level of perceptual organization. And it is connected with simultaneous and global processing of information based upon visual perception, which enables integration of perceived elements and combining them into meaningful wholes. The rate of mental processes as well as eye-hand coordination were also good. They might have difficulties due to the necessity to work under time pressure and their weak resistance to distracters. On the other hand, all of them scored very low on verbal perception tasks, and exhibited difficulties with comprehension of verbal commands and instructions.

It does not need reminding that communication impairment is the main diagnostic feature of autistic triad. Yet, the majority of studies relay heavily upon verbal instructions given to the examined children. The authors seem to forget that autistic children may have difficulties with understanding verbal commands or need some time to process the information included there. Hence, if the scores of sequential processing of abstract nonverbal concepts are within average range, it allows the conclusion that a given child is also able to perform logical operations on verbal material. Especially, if she is able to see cause and effect relationships in everyday situations. Beside communication problems difficulties with performing experimental tasks may also be due to the necessity of combining two types of clues: verbal (instruction) and nonverbal (demonstration). So the actions of an examiner, who intends to explain the rules of a given task, may in effect lead to a confusion as was the case with well known Piaget's experiments (see Donaldson, 1978). It is also reflected in the phenomenon of "horizontal decalâge". That is an inconsistency in the tasks healthy children can perform. For instance, they can solve Piaget's conservation problem on numbers at age six, but they are not able to solve it on mass until age eight, while the ability to perform conservation of weight task appears only at the age of ten (Wortman, Loftus and Marshall, 1988).

Bearing all this in mind we have used a number of repetitions in order to make sure that the child had understood what was required from her in a given experimental situation. The aim of the above described task was to make it the child understand the visual perspective of another person. While showing her work to the parent the child established a common field of sharing attention. Awareness of the existence of two different points of view is an important indicator of developing social interactions (Meltzoff, 1995; Repacholi & Gopnik, 1997; Gopnik, Meltzoff & Kuhl 1997). A sequence of behaviours used in the above described experiment was to create an awareness of different points of view in autistic children. As pointed out above the girls were much better in performing those tasks than the boys we were examining during a ten year period.

8.2 Possible sources of differences between girls and boys

It is often stressed in autistic literature that the symptoms of ASD exhibited by girls are more difficult to diagnose than those occurring in boys (Attwood, 2007; Kopp et al., 2010; Skuse, 2009). One of the reasons is higher than in boys' level of social skills, such as an ability to come into social interactions and a style of behaviour in general. Moreover, deviant behaviours of children with ASD influence not only their own development but also the life of the whole family. At the same time, they have impact upon their relations with peers and adults from outside the family. It concerns their attitudes towards such children in particular since autistic children were believed - and often still are – to be dangerous for others due to their bizarre and odd behaviours. Since boys are generally believed to be more aggressive, another significant factor discriminating functioning of girls and boys with ASD may be a difference in educational treatment (Constantino et al., 2009: Jonson-Reid et al., 2010). Beside social learning biological factors may also play an important role. One of them is the testosterone impact upon the foetus (Baron-Cohen et al., 2005).

It may also be worth to point to another factor that might cause differences between symptoms observed in boys and girls. Namely, a well known clinical fact that female brain is less localized than the male brain (Moir and Jessel, 1992). In consequence, brain lesions in women are less disastrous than in men what may be best observed in aphasia recovery. As pointed out earlier, the process of brain development is connected with elimination of unnecessary connections, which leads to refinement of behaviour (see also Kaczmarek, Markiewicz, 2008). It is highly plausible that in the case of autism the selection is delayed, which results in over-abundance of connections, and in disinhibition leading to sensory overload.

It was noted in other works (Kaczmarek, 2003; Kaczmarek, Markiewicz, 2008) that while creating our own image of the world we concentrate upon matters that are important for us and leave out less significant. Therefore, our world image is simplified to a considerable degree, and thanks to it the surroundings seem predictable. And it is that presumed predictability that gives us a feeling of safety. Thus, meeting a mentally ill person makes us feel a bit nervous because we do not know what to expect from him. The autistic child is not able to single out significant stimuli from the non-significant ones, hence her world becomes unpredictable, incomprehensible, and terrifying. It may lead not only to fits of aggression and self-aggression but also to stereotyped repetitive behaviours so characteristic of autism.

9. Conclusion

Our own clinical practice as well as other studies show that early diagnosis and therapeutic procedures connected with it facilitate socialization of children with ASD. Of particular

significance is the individual approach to each autistic child. It enables evaluating not only level of actual skills but also developmental potential of a given child, which in turn improves efficacy of treatment. If we know a zone of proximal development of a given child, we are able to concentrate on the areas in which the child is prone to succeed. Therefore, the microgenetic analysis of the manners in which the child strides to overcome difficulties while solving particular tasks proves to be of great significance. It gives a therapist tools for developing potential abilities of the child, and not to concentrate on her disabilities as it is often the case. Such an individualized and progressive approach increases the efficacy of therapy and gives the child a feeling of success stimulating her to work.
Taking into account developmental profile of a particular child is a necessity since there is a considerable differentiation among autistic persons both in the inter- and intra-individual dimensions. One of them is the difference between the clinical picture of symptoms in boys and girls. Moreover, studies of Constantino and collaborators (2009) revealed subtle difficulties in communicating with others in 20 per cent of families of children with ASD. They were observed mainly in siblings in whom some traits of autism appear about ten times more often than in healthy population. The authors are of the opinion that at least some of specific traits of autistic spectrum may be hereditary. Therefore, they stress the necessity of taking into consideration sex differences while making a diagnosis of autism.
Our study has shown that it is possible to develop a schema of action, or a script, in autistic girls. Moreover, the techniques we applied made it possible to analyze a given change, while it was actually happening, and not comparing behavioral patterns from a pre- and post-change as it is often done.Yet, the script remains a rigid unchangeable sequence of actions, while healthy people change it in accordance with the requirements of environment. It was also noted that the autistic symptoms in girls are less pronounced due mainly to their better communicative competence. For that reason their disorders are often neglected despite the fact that their qualitative character is not much different from the disturbances observed in boys. The differences are mainly of a quantitative nature. It is quite probable that refinement of diagnostic methods will lead to better understanding of their problems causing a dramatic increase of the number of autistic girls as was the case with other clinical syndromes.

10. References

Attwood, T. (2007). *The Complete Guide to Asperger's Syndrome,* Jessica Kingsley Publishers, ISBN 1-84310-495-4, London,England.

Auyeung, B.; Baron-Cohen, S.; Chapman, E.; Knickmeyer, R.; Taylor, K. & Hackett, G. (2009). Foetal Testosterone and Autistic Traits. *British Journal of Psychology,* Vol. 100, pp. 1-22, Online ISSN 2044-8260

Baron-Cohen, S.; Knickmeyer, R. & Belmonte, M. (2005). Sex Differences in the Brain: Iplications for Explaining Autism. *Science,* Vol. 310, pp. 819-823, Online ISSN 1095-9203

Brown, J.W. (1988) *The Life of the Mind.* ISBN 0-8058-0236-3, LEA, Hillsdale.

Brown, J.W. (2001). Microgenetic Theory: Reflections and Prospects. *Neuropsychoanalysis,* Vol. 3, pp. 61-74, ISSN 1529-4145

Brown, J.W. & Pachalska, M. (2003). The Nature of the Symptom and Its Relevance for Neuropsychology. *Acta Neuropsychologica,* Vol.1, No.1, pp. 1-11, ISSN 1730-7503

Calais, G.J. (2008). Microgenetic Analysis of Learning: Measuring Change as It Occurs. *National Forum of Applied Educational Research Journal*, Vol.21, No. 3 (24 February 2011) Available from http://www.nationalforum.com/ElectronicJournalVolumes/Calais,Gerald Microgenetic Analysis Learning.pdf

Chakrabarti, B.; Dudridge, F.; Kent, L.; Wheelwright, S.; Hill-Cawthorne, G.; Allison, C.; Banerjee-Basu, S. & Baron-Cohen, S., (2009). Genes Related to Sex-steroids, Neural Growth and Social-emotional Behaviour are Associated with Autistic Traits, Empathy and Asperger Syndrome. *Autism Research*, (24 February 2011) Available from http://onlinelibrary.wiley.com/doi/10.1002/aur.80/pdf

Constantino, J.N.; Abbacchi, A.M.; Lavesser, P.D.; Reed, H.; Givens, L.; Chiang, L.; Gray, T.; Gross, M.; Hang, Y. & Todd, R.D. (2009). Developmental Course of Autistic Social Impairment In males. *Development and Psychopathology*, Vol. 21. No.1, p.p. 127-38, ISSN 1469-2198

Donaldson, M. (1978). *Children's Minds*. Fontana/Collins, ISBN 0-85664-759-4, Glasgow, Scotland

Flynn, E.; Pine, K. & Lewis, C. (2006). The microgenetic method: Time for change? *The Psychologist*, Vol.19, No.3, pp. 152-155, ISSN 0952-8229

Geschwind, N. & Galaburda, A. M. (1985a). Cerebral Lateralization. Biological Mechanisms, Associations, and Pathology. I. A Hypothesis and a Program for Research. *Archive of Neurology*, Vol. 42, pp. 428–459, Online ISSN 0375-8540

Geschwind, N. & Galaburda, A. M. (1985b). Cerebral Lateralisation: Biological Mechanisms, Associations and Pathology. II. A Hypothesis and a Program for Research. *Archives of neurology*, Vol. 42, pp. 521–552, Online ISSN 0375-8540

Geschwind, N. & Galaburda, A. M. (1985b). Cerebral Lateralisation: Biological Mechanisms, Associations and Pathology. III. A Hypothesis and a Program for Research. *Archives of Neurology*, Vol. 42, pp. 634–654, Online ISSN 0375-8540

Gopnik, A.; Meltzoff, A.N & Kuhl, P.K. (1999). *The Scientist in the Crib* . Hapner Collins, ISBN 0-688-15988-5, New York,U.S.A.

Jonson-Reid, M.; Presnall, N.; Drake, B.; Fox, L.; Bierut, L.; Reich, W.; Kane, P.; Todd, R.D. & Constantino, J.N. (2010). Effects of Child Maltreatment and Inherited Liability on Antisocial Development: an Official Records Study. *Journal of the American Academy of Child and Adolescent Psychiatry*, Vol. 49, No.4, pp. 321-332, Online ISSN 1527-5418

Kaczmarek, B.L.J. (2003). The Life of the Brain. *Acta Neuropsychologica* Vol.1, No.1, pp. 12- 21, ISSN 1730-7503

Kaczmarek, B.L.J. & Markiewicz, K. (2008). The Brain and the Mind, In: *Neuropsychology and Philosophy of Mind in Process. Essays in Honor of Jason W. Brown*, M. Pąchalska & M. Weber (Eds.), 98 - 109, Ontos Verlag,ISBN 978-3-86838-110-1, Frankfurt, Germany

Kopp, S. (2010). Girls with Social and/or Attention Impairments. (12 February, 2011) Available at http://gupea.ub.gu.se/bitstream/2077/23134/3/gupea_2077_23134_3.pdf

Kopp, S.; Beckung, E. & Gillberg, C. (2010). Developmental Coordination Disorder and other Motor Control Problems in girls with Autism Spectrum Disorders and/or Attention-deficit/Hyperactivity Disorder. *Research in Developmental Disabilities*, Vol. 31, pp. 350-361, ISSN 0891-4222

Lord, C. & Schopler, E. (1985). Differences in Sex Ratios in Autism as a Function of Measured Intelligence. *Journal of Autism and Developmental Disorders*, Vol.15, pp. 185-193, Online ISSN 1573-3432.

Lord, C.; Schopler & E., Revicki, D. (1982). Sex Differences in Autism. *Journal of Autism and Developmental Disorders*, Vol.12, pp. 317-330, Online ISSN 1573-3432.

Markiewicz, K. (2008). Disturbances of Central Coherence. A Case Study. *Acta Neuropsychologica*, Vol. 6, No.2, pp. 83-93 , ISSN 1730-7503

Markiewicz, K. (2009). Disturbances of Executive Function in Autistic Children. *The New Educational Review*, Vo.17, No. 1, pp. 361-366, ISSN 1732-6729

Markiewicz, K. & Pąchalska, M. (2007). Diagnosis of Severe Developmental Disorders in Children under Three Years of Age. *Medical Science Monitor*, Vol.13, No.2, pp. 89-99, Online ISSN 1643-3750

Markiewicz, K. & Grochmal-Bach, B. (2008). Correlation of the Theory of Mind and the Level of Cognitive Functioning in Autistic Children. *Acta Neuropsychologica*, Vol.6, No.3, pp. 247-252, ISSN 1730-7503

Markiewicz K., MacQueen B.D. (2008). The Autistic Mind: a Case Study. *Medical Science Monitor*, Vol.15, No.1, pp. 5-13, Online ISSN 1643-3750

Markiewicz, K.; Kaczmarek, B.L.J.; Grochmal-Bach, B.; Olszewski, H. & Guzińska, K. (2009). Rehebilitation of Executive Functioning in an Autistic Child: A Case Study. *Acta Neuropsychologica*, Vol.7, No.1, pp. 45-58, ISSN 1730-7503

McLennan, J. D.; Lord & C., Schopler, E. (1993). Sex Differences in Higher Functioning People with Autism. *Journal of Autism and Developmental Disorders*, Vol. 23, p.p. 217-227, Online ISSN 1573-3432

Meltzoff, A.N. (1995). Understanding the Intentions of Others: Reenactment of Intended Acts by 18-month-old children. *Developmental Psychology*, Vol.31, pp. 838 - 850, Online ISSN 1939-0599

Newschaffer, C.J.; Crown, L.A.; Daniels, J.; Giarelli, E.; Grether, J.K.; Levy, S.E.; Mandel, D.S.; Miller, L.A.; Pinto-Martin, J.; Reaven, J.; Reynolds, A.M.; Rice, C.E.; Schendel D. & Windham, G.C. (2007). The Epidemiology of Autism Spectrum Disorders. *Annual Review of Public Health*, Vol.28, pp. 235–58, ISSN 0163-7525

Moir, A. & Jessel, D. (1991). *Brain Sex: The Real Difference Between Men and Women*. ISBN 0-385-31183-4, Dell Publishing, New York, U.S.A..

Nichols, S. (2009). A Girl's-Eye View: Detecting and Understanding Autism Spectrum Disorders in Females (12 February 2011) Available from http://www.iancommunity.org/cs/articles/.

Nichols, S.; Moravcik, G. & Tetenbaum, S. P. (2008). *Girls Growing up on the Autism Spectrum.: What Parents and Professionals Should Know about the Pre-teen and Teenage Years*. Jessica Kingsley Publishers, ISBN 9781843108559, London, England

Repacholi, B.M. & Gopnik, A. (1997). Early Reasoning about Desires: Evidence from 14- and 18-month-olds. *Developmental Psychology*, Vol.33, pp. 12 -21, Online ISSN 1939-0599

Siegler, R.S. & Svetina, M. (2002). A microgenetic/Cross - Sectional Study of Matrix Completion: Comparing Short - Term and Long - Term Change. *Child Development*, Vol.73, No.3, pp. 793 - 809, Online ISSN 1467-8624

Skuse, D.H. (2009). Is Autism Really a Coherent Syndrome in Boys, or Girls? *British Journal of Psychology*, Vol.100, No.1, pp. 33–37, Online ISSN 2044-8295

Tsatsanis, K.D. (2004). Heterogeneity in Learning Style in Asperger Syndrome and High-functioning Autism. *Topics in Language Disorders,* Vol. 24, No.4, pp. 260-27, ISSN 0271-8294

Wagoner, B. (2009). The Experimental Methodology of Constructive Microgenesis. In: *Dynamic Process Methodology and Developmental Sciences,* J. Valsiner; Molenaar, P.C.M.; Lyra, M.C.D.P.; Chaudhary, N. (Eds.), 99-121, e-ISBN 978-0-387-95922-1, Springer, Dordrecht, The Netherlands, (18 April 2011) Available from http://aalborg.academia.edu/Brady Wagoner.

Wortman, C.B.; Loftus, E.F. & Marshal, M.E. (1988). *Psychology.* Alfred A. Knopf, ISBN 0-394-36537-2, New York, U.S.A.

10

Genetic Counseling in Autistic Phenotypes

Agnes Cristina Fett-Conte
Medical School - FAMERP/FUNFARME, São José do Rio Preto, São Paulo
Brazil

1. Introduction

Autism is a neurobehavioural disorder that includes impairment in social interaction and language development and communication deficits accompanied by repetitive and stereotyped behaviours. More recently this term has been used to define a very broad behavioural phenotype which is classified as different disorders that comprise the Pervasive Developmental Disorders (PDD) according to the Diagnostic and Statistical Manual of Mental Disorders, 4th Edition-DSM-IV (American Psychiatric Association [APA], 1994). It contains the criteria for diagnosis and specific characteristics of each disease, including Autism, Asperger's syndrome, Childhood Disintegrative Disorder, Rett syndrome and Pervasive Developmental Disorder Not Otherwise Specified (PDD-NOS).

However, with the exception of Rett syndrome, the others make up a continuous spectrum rather than clinically defined diagnostic categories due to the wide variation of clinical signs and symptoms and the subjectivity of the criteria for differential diagnosis. For this reason these disorders have been included in a general conceptual category, Autism Spectrum Disorders (ASDs) (Snow & Lecavalier, 2011; Witwer & Lecavalier, 2008). Hence, the proposals for DSM-V, being prepared by the APA, which is scheduled to be published in 2012 or 2013, recommend that Rett syndrome is not considered among the ASDs, that the designation PDD is no longer used and that ASD is considered a single category that includes Autism, Asperger's syndrome, Childhood Disintegrative Disorder and PDD-NOS. That is, the disorders that compose the autistic spectrum would no longer have specific names (APA, 2011).

Rett syndrome almost exclusively affects girls and is characterized by normal development until about six months followed by regression of motor and social skills. The triad dementia-ataxia-autism is observed as is a characteristic pattern of deceleration in the rate of head growth, loss of acquired manual skills, poorly coordinated gait, involuntary movements of the hands and the trunk and autistic features. Epilepsy may be present and a abnormal respiratory pattern is typical . The prevalence among women is between 1:10,000 and 1:15,000 with most cases caused by a sporadic mutation in the *MECP2* gene, located on Xq28. In some cases, the etiology is due to mitochondrial DNA mutations (Gonzales & LaSalle, 2010; Nissenkorn et al., 2010; Temudo et al., 2010). The peculiar nature and specific etiology, linked to a genetic defect with consequent brain damage, are among the reasons for not being considered within the ASDs.

Childhood Disintegrative Disorder is basically characterized by normal development of children until at least two years of age followed by a process of loss of previously acquired intellectual and behavioural skills, which results in autistic behaviour (Homan et al., 2011).

Asperger's syndrome differs from other diagnoses because of the absence of delay in language development, in general a preserved cognitive development, frequently prodigious memory, as well as "pedantic" speech, inadequate social interaction and, in many cases, disinterest in interpersonal relationships (Koyama & Kurita, 2008).

Since the reports of Kanner in 1943 on "autistic disturbances of affective contact", Autism has been extensively discussed and investigated. Currently regarded as a developmental disorder that manifests before thirty months of age, it is characterized by abnormal responses to auditory and visual stimuli and underdeveloped or absent speech. Serious communication and social interaction problems occur and behaviour is ritualistic, aggregating abnormal routines with resistance to change. Approximately 75% of cases are associated with mental retardation, 15 to 40% with seizures and 20 to 50% with electroencephalographic abnormalities (Tuchman et al., 2010).

PDD-NOS is a diagnosis of exclusion made when an individual presents severe impairment of reciprocal social interaction and verbal or nonverbal communication skill development, but does not satisfy the criteria for other PDDs. Atypical Autism, for which the etiology and prevalence remain unknown, is also included in this category (Chiappedi et al., 2010; Koyama & Kurita, 2008).

ASDs occur in approximately 1:150 live births and in all ethnic groups and social classes, and thus can be considered a public health problem. There are discussions as to whether there is a real progressive increase in the prevalence of these diseases in the population and it is speculated that there are several risk factors. However, this increase appears to result from the fact that diagnosis is being made earlier as education and healthcare staff are more attentive to the symptoms, besides the diffusion of information, leading to the identification of a greater number of cases (Liu et al., 2010; Shen et al., 2010; Nassar et al., 2009). There are several diagnostic scales that use "checklists" and are effective in the rapid identification of possible cases of ASD. The degree of behavioral and cognitive functioning is highly variable and early diagnosis is of paramount importance because stimulation programs achieve much more significant results when interventions occur in the early development stages. If diagnosis and intervention are delayed, the results are not very promising (Biederman et al., 2010; Marteleto et al., 2008).

But if the classification of the autism phenotype is so difficult and so discussed, the etiology is even more so. Knowledge about the etiology of ASDs is increasing, but causes remain elusive for most cases. The truth is that autism has many etiologies.

ASD associated with a known cause is called syndromic autism. There is an expanding list of medical conditions in the literature associated with autistic manifestations, ranging from disruptions caused by varying environmental agents to several mutations and well-defined syndromes, chromosomal abnormalities and metabolic diseases. In cases where the cause is identified, the autistic manifestation is considered secondary (Benvenuto et al., 2009). Among these, prenatal infections, prenatal exposure to physical and chemical agents and genetic disorders may be cited (Ratajczak, 2011; Zhang et al., 2010). However, the biological mechanisms involved in these associations are unclear.

The clinical heterogeneity of ASDs probably reflects the complexity of the genetic profile. There is no doubt that different genetic mechanisms contribute to the pathogenesis of ASDs. Thus, when the many different etiologies of autistic phenotype are referred to, the principal focus is on genetic aspects. Heritability is estimated in 90% and the monozygotic twin concordance rate is as high as 95%. The situation is complicated by significant inter-individual heterogeneity, the numerous loci involved and gene-environment interactions (Caglayan, 2010).

Another interesting aspect is related to the phenomenon of genetic anticipation. Since the first descriptions by Kanner, particular personality traits in relatives of autistic patients have been recognized. The findings of familial aggregation of minor variants suggest that genes confer susceptibility at variable severity, which is often "light", known as *broad phenotype*, and independently segregates among relatives (Losh et al., 2008; Schmidt et al., 2008).

This suggests that this complex combination of genetic and environmental factors, is what really defines the risk for ASDs. The commonly accepted empirical risk estimate for a couple with one affected child is 2–8%, in the absence of a definablecondition (Selkirk et al., 2009).

Karyotype analysis shows changes involving all chromosomes in 3 to 6% of ASD cases. However, the functional significance of these changes also remains unknown given the variation in the size of the regions involved and the diversity of loci. Moreover, the majority of rearrangements are sporadic, some are detected in other asymptomatic family members or are *de novo* in individuals with a positive family history of ASDs (Marshall et al., 2008; Sykes & Lamp, 2007).

Many genes are likely to contribute to the etiology of ASDs, especially in cases of non-syndromic autism, as they present mutations or polymorphisms. The identification is becoming easier as a result of advances in genetic technology. It is believed that the emergence of the autistic phenotype in most cases depends on a small additive effect of multiple genes, but all with expressions in the central nervous system. Among these are the *CENTG2* gene mapped at 2q37.2, the *SHANK3* gene mapped at 22q13.3, the *GABRB3* gene mapped at 15q11-13, the *SLC6A4* gene located at 17q11.2 and the *NLGN3* gene mapped at Xq13.1 (Cuscó et al., 2009).

Many studies recommend that the laboratory evaluation of ASD cases should initially include an analysis of G banded karyotype, preferably high resolution and a molecular evaluation of the *FMR1* gene. But even at high resolution, abnormalities smaller than ~5Mb cannot be detected by karyotyping, which is problematic, particularly in subtelomeric chromosomal regions that are rich in genes susceptible to rearrangements. For this reason, karyotypic evaluation by the Fluorescent in situ Hybridization (FISH) technique is indicated to overcome some limitations and clarify certain karyotypic findings. However, negative results obtained with these techniques have not ruled out other types of genetic alterations.

Genetic screens represent a powerful tool when dealing with monogenic disorders characterized by direct genotype-phenotype correlations. Current guidelines for clinical genetic evaluation of patients recommends carrying out a detailed physical examination, hearing evaluation, obtaining a detailed personal and family history, screening for inborn errors of metabolism and neuroimaging studies, as well as karyotype and fragile-X DNA testing (Lintas & Persico, 2008; Wassink et al., 2007). The identification of genes linked to susceptibility and investigation of pathogenic mechanisms is crucial in clinical practice and for adequate genetic counseling of families, but the specifications and limitations of each test should be considered. Some genetic testing, even as part of research protocols for ASDs, can only be time consuming and not appropriate in many cases.

More recently tests to identify cryptic genomic changes have been proposed. The development of array-based CGH (Comparative Genomic Hybridization) and MLPA analysis (Multiplex Ligation-dependent Probe Amplification) has enabled detection of microdeletions and microduplications in patients with ASDs. These have been referred to as copy number variants (CNVs) and seem to play a key role in the etiology of many cases, more commonly among patients with non-dysmorphic ASDs (Benvenuto et al., 2009; Christian et al., 2008). But despite the promising genetic findings, the data are still inconclusive which is due to genetic heterogeneity, the likely involvement of many genes that interact, epistatic interactions, gene-environment interactions, variability in gene expression, the influence of epigenetic mechanisms and the fact that the expression of some genes is influenced by specific regulatory regions located at relatively long distances, even on other chromosomes, which makes the selection of candidate genes difficult (Zahir & Brown, 2011; Vorstman et al., 2006). The fact that the cost of these tests is high and the availability is low has to be considered as this makes access for many patients difficult.

The high prevalence and complexity of the ASDs have motivated several studies using different research strategies. Genetic factors are the most studied and its potential cause in many cases has resulted in a significant increase in the number of referrals to clinical geneticists and genetic counselors.

Genetic counselors are able to help families that have children with syndromic autism and even in cases with uncertainty regarding etiology. But, genetic counseling for families of ASD individuals is a difficult procedure. The most important aspect is that genetic counseling is not only a question of giving technical information related to all the complexity of the aforementioned features. Even so, technical information can be offered in several contexts such as healthcare and educational booklets and even in television shows.

Genetic counseling is the process of providing information to individuals and families about the nature, inheritance, and implications of genetic disorders to help them make informed medical and personal decisions. It is a communication process. As such it should be understood as a "two-way street", i.e. as a situation of "exchange". Counselors have no guarantee or control that their "message" to counselees is understood as intended, nor even about the consequences of the process. Thus, besides the communication of biological and clinical information, counselors must prioritize the educational and psychological aspects of the process, so that all the information and emotional support given to counselees can support their decision making and help to reduce anxiety and guilt. It is essential to remember that nondirectiveness is crucial in this process (Kessler, 2001) as is the context, the environment chosen for the process to develop.

Nondirectiveness is not a question of whether to give advice or not or to say what the counselor thinks is best or not. It is a form to promote and to enhance the autonomy and self-directedness of counselees. It is necessary to provide accurate, complete and unbiased information and to have an empathic relationship between those involved, professionals and family. Nondirectiveness and ethical principles applied to genetic counseling are very well documented in a publication of the World Health Organization in 1998 (World Health Organization [WHO], 1998).

There are different methods to promote the identification of the most relevant aspects that must be addressed with the families. The counselor should convey all the useful information requested, as well as information that should have been requested by counselees, but was not.

All circumstances of genetic counseling are in a complex context that involves personal dynamics and social interactions with the meaning and perception being very different among those involved (i.e., counselees and counsellor). Invariably, however, stress and anxiety are present in this context. It should be noted that coping strategies differ between individuals, ranging from seeking information or avoiding new information to reactions of anger or indifference, which are psychological defences against an aversive event that causes pain. Often, the cascade of psychological effects that begins is unpredictable. Certainly, at least in the first session of genetic counseling, the counselor is faced by shocked and very vulnerable individuals with a great sense of loss, guilt and shame. Family issues, structure, emotional dynamics, religion, patterns of communication, kinds of interactions, ethnicity and social support must be considered during the counseling process because all these issues will influence the counseling.

The diagnosis of autism is really a major stressor for families who have to adapt to a reality that, in addition to being new, is very heterogeneous, complex, and difficult and can result in conflict. This requires professionals working in this area to invest more in psychosocial genetic counseling skills. These issues and their applications to genetic counseling were detailed by Weil (2000).

Genetic counseling generally involves a chronicity situation which is absolutely true in the case of autism. Chronic diseases can produce consequences such as pain, discomfort, low self-esteem, uncertainty about the future, suicidal thoughts, fear, panic, general and specific disorders of conduct, academic performance deficits, difficulties in interpersonal and family relationships, anxiety, and depression among others. The emotional distress associated with these diseases, if ignored, can lead to a significant reduction in the quality of life of patients and their families and negatively affect the absorption of important information and adherence to treatment. Thus, family members should always be considered at risk for developing some kind of emotional disorder. These considerations underscore the nature of genetic counseling as something far beyond the process of medical diagnosis and the establishment of the risk of occurrence/recurrence. Hence, skilled and experienced professionals are needed to perform this task, giving priority to communications and humanization of care. In the past, communication skills were not considered a priority. However, today these skills have become a professional demand and even the legal obligation of every professional in healthcare.

Genetic counseling is developed in a continuous and integrated manner. Division in phases is only for teaching purposes and can be summarized as: the reception and identification of patient/family, understanding of the problem/complaint, the identification of antecedents, establishment and confirmation of the diagnosis, assessment of genetic risk, discussion about options and decisions, and follow up. Psychological support from the counselor is essential for each phase, whether in a single session or several.

There are not a great number of reports about genetic counseling in ASDs. Maybe one of the reasons is related to the misunderstanding about the heritability of these disorders as mentioned above. Like other diseases, some cases are inherited and others are not. Then, what to do in each case since genetic causes may play an important role in the etiology of ASDs?

The clinicians have to identify specific causes or exclude them to provide effective counseling and this is not always an easy job. Two situations must always be considered: syndromic autism and non-syndromic autism (idiopathic or primary). In the first one, there is a known cause related to the behavioural phenotype and that often can be identified by dysmorphic features. It may be associated with well-known monogenic disorders, chromosomal alterations and environmental events. Genetic counseling should be directed to information related to the cause, genetic or not. Someone could say that if the cause is not genetic, the patient certainly will not go to a genetic counseling service, but in practice this is not true. In non-syndromic cases, determined after detailed investigations, the approach will be different, discussing in particular the polygenic predisposition and the environmental contribution to the autistic phenotype.

2. Genetic counseling in genetic disease associated to autism

Approximately 10% of autistic patients have a diagnosis of single gene diseases or chromosomal abnormalities; there are several molecular pathways potentially involved in the alterations that affect normal neurodevelopmental events. In cases of chromosomal defects, for example, these can cause alterations in neuronal migration and brain growth, with subsequent altered cortical organization, synaptic and dendritic changes and the ASD phenotype. Metabolic disorders produce an accumulation of toxic metabolites which can cause a reduction of myelin, neuronal loss, alterations in dopaminergic or serotonergic neurotransmission and ASD (Benvenuto et al., 2009).

The most common genetic diseases associated with ASDs include Fragile-X syndrome, Tuberous Sclerosis, invdup(15) or idic(15), Prader-Willi and Angelman syndromes, Down syndrome, Joubert syndrome, macrocephaly and overground syndromes, Turner syndrome, Williams syndrome, Timothy syndrome, Smith-Magenis syndrome, Phelan-McDermid (22q13.3 deletion) syndrome, Cohen syndrome, Sanfilippo syndrome, mitochondrial cytopathies, among others (Caglayan, 2010; Moss & Howlin, 2009). And this list just keeps growing!

A Fragile X syndrome is diagnosed in almost 5% of the children with ASDs. Faced with this, it is considered the most common genetic etiology of the autistic phenotype. It is the most common cause of inherited intellectual deficiency in men. It results from a full mutation that affects approximately 1 in 2500 males and 1 in 8000 females. The molecular basis involves a dynamic and unstable mutation characterized by the repeat expansion of the trinucleotide $(CGG)_n$ in the 5′ untranslated region of the first exon of the *FMR-1* gene (Fragile-X Mental Retardation 1) mapped at Xq27.3. When the number of CGG repeats is greater than 200, the allele is classified as a full mutation. CpG island hypermethylation of the promoter causes gene inactivation. Persons with the syndrome produce little or no detectable expression of the encoded protein called Fragile X Mental Retardation Protein or FMRP which is essential for normal brain function. It is involved in synaptic maturation and its loss may alter neuronal plasticity. Brain damage results in, among other things, autistic behaviour. Individuals with intermediate CGG expansions in the range of 55–200 repeats are known as fragile X premutation carriers and are at increased risk for a related disorder known as Fragile X-Associated Tremor and Ataxia Syndrome (FXTAS) that affects primarily men over the age of 50. The presence of the premutation in women can also cause premature ovarian failure (Hampson et al, 2011).

All the diseases associated with autism have specific molecular biological mechanisms with different genes involved and different types of inheritance pattern. The fact is that the autism phenotype is one of the clinical manifestations of the disease itself, which in one way or another, changes the structure and/or function of the brain. For this reason, it is spoken of as association rather than comorbidity because the events are not random in the same patient. Given this scenario, genetic counseling should be directed according to information relevant to that specific syndromic diagnosis with explanations of the causes, risks and consequences. Conduct is not so differentwhen the etiology of autism arises from the action of a toxic environmental agent.

The autism phenotype, however, is a major complicating factor when combined with a genetic disease because the parents, who are usually very distressed, can confuse the risk of disease recurrence with risk of autism among those affected by it. Not all of those affected have the autism phenotype. It is important that counselors, in addition to background and needs, identify the expectations of their genetic counseling clients. However, this scenario includes hundreds of possibilities of events and different strategies to solve them. Some of the genetic conditions are inherited while others are not. In most cases there is an important variability in clinical manifestations. If the disease is genetic it is incurable although often there may be symptomatic and palliative treatments.

Autism phenotype can be associated with autosomal recessive disease, which was originated from parents who carry the deleterious gene. This is one of situations of genetic counseling that inherently evoke guilt. The dominant culture of the family, especially if its members are Latino, produces the feeling of being punished for some sin. In this case the guilt can be a response to new and adverse reality over which one has no control. This can be exacerbated if during the explanation the counselor emphasize features such as that the probability of the outcome was very low.

Sometimes it is difficult for parents to 'see' the genetic disease of their children since the autistic symptoms appear more strongly than the dysmorphic features. By a lack of standardized diagnostic procedures in many syndromes and the absence of laboratory markers, the diagnostic process often stems from interpretation of a set of clinical signs and the experience of the geneticist. For some families, accustomed to different clinical procedures, this can also cause anxiety.

In some cases there is a probable diagnostic hypothesis, however, the test(s) required to arrive at an accurate diagnosis may not be accessible to the family. This can greatly hinder the process of genetic counseling and create stress in the family and counselor. The molecular revolution observed in the last three decades has introduced many procedures that are not still available in public health programs of several countries and only the most economically advantaged families can access them, which does not correspond with the reality of most people. While it lasts, intercountry collaboration programmes should be stimulated (WHO, 2010).

A large number of families consult the Internet before the counseling to obtain information about the diagnoses, treatment, and tests and so many clients arrive for genetic counseling with notions of the condition for which they are to have counseling (Peters & Petrill, 2011). This creates a series of expectations. Not always, however, information is obtained from reliable sources and it is for the counselor to clarify false beliefs or possible misinterpretation.

Also, it must be emphasized that the search for a solution makes the Internet a tool that frequently causes more harm than good. The demand for treatment has increased gradually and cognitive behavioural intervention programmes aimed at trying to improve social interaction and communications are encouraged (Wood et al., 2009). The design of these interventions is to act during the critical period of postnatal neuronal plasticity (within the first three years of life). But there are other not empirically proven therapies; for this reason, sites selling solutions for autism have proliferated. Couples come to genetic counseling requesting an opinion and explanation from the counselor on "magic formulas";they become anguished and even feel guilty when they realize that this solution is unfeasible, especially as some of them have very high costs. It is for the counselor to reduce the anxiety of parents and explain that this is not about being for or against any type of alternative therapy, but that most have no scientific basis and some may even pose health hazards. Families need to understand the evidence for efficacy (or lack thereof) and potential side effects. More accurate and earlier diagnosis or the elucidation of etiological factors does not mean effective therapies in the short term.

3. Genetic counseling in autism of unknown etiology

On taking into account all technologies, an underlying genetic diagnosis is identified in around 10–15% of ASDs cases while cytogenetically visible chromosomal rearrangements are found in 2-6% of ASDs individuals (Bremer et al, 2011; Kumar & Christian, 2009). Hence, for most individuals (90%) with the autistic phenotype, there is no known genetic or environmental cause, which defines them as non-syndromic or "idiopathic' as previously mentioned. Often this condition is established after negative resultsobtained from a medical evaluation to identify medical issues that affect the development and behavior of nonverbal children, physical examinationabout metabolic, medical, or neurologic conditions, careful examinationof personal history, a detailed investigation of gestational antecedents and dysmorphic signs and after performing an odysseyof multiple testing.

Genome-wide studies have implicated numerous minor risk alleles with low and high penetrance but few common variants and with many contributing loci. Among the candidates are genes that code for important proteins in synaptic structure, function and maintenance. Genetic mutations in these genes result in an aberrant synaptic process that could produce the ASDs phenotypes. However, the frequency of these mutations is so low that widespread screening does not seem to be clinically justified. Some, however, deserve to be investigated because of clinical findings such as mutations in the *PTEN* gene in children with macrocephaly (Lintas & Persico, 2008).

As etiological factors are progressively being discovered, it is natural to think that the number of idiopathic cases will also gradually decrease. The increased resolution of CGH array testing in combination with new technologies, such as whole genome sequencing and bioinformatics programs, will play an important role in helping us to further understand the complex genetic basis of autism. The implementation of these high resolution techniques in the genetic research of ASDs may discover specific genotypes and subtypes of ASDs for which new diagnostic and therapeutic strategies can be developed. For this reason the identification of genetic abnormalities is a high priority in the study of ASD (Bremer et al., 2011).

For now, non-syndromic cases are much more common than other forms with estimates in the general population reported at approximately 1 in 100. In these cases, ASD is considered a complex disease of multifactorial pattern inheritance (Harrington, 2010; Maenner & Durkin, 2010)[4] M.J. Maenner and M.S. Durkin, Trends in the prevalence of autism on the basis of special education data, Pediatrics 126 (2010), pp. e1018–e1025. Full Text via CrossRef | View Record in Scopus | Cited By in Scopus (1). About 70% of probands with autism of unknown cause has a first- or second-degree relative with autistic symptoms, and 15% has fathers with Asperger syndrome. The empiric aggregate risk to sibs of individuals with autism of unknown cause varies across studies but is generally considered to range from 5% to 10% for autism and 10% to 15% for milder symptoms, including language, social, and psychiatric disorders. For families with two or more affected children, the recurrence risk approaches 35% (Miles et al., 2010).

All this information should be thoroughly discussed with the members of the family at their level of understanding. Obviously, faced with such uncertainty and heterogeneity, the counselor may feel uncomfortable to report these risks. It is essential that the family understands that when a child is diagnosed with an ASD, a range of etiological options are involved, which means the possibility of many different diseases. An aggravating factor is that the information may generate anxiety; most families have social and institutional barriers to carrying out more sophisticated tests.

4. Psychosocial aspects of genetic counseling in autism

Throughout its development, the family goes through many changes. Each phase of the so-called life cycle (acquisition, adolescence, maturity and final) has its own peculiarities and difficulties inherent to the transformations that occur. During the acquisition phase, with the arrival of children, accepting parenting is already difficult. The family system "grows" as a whole and new links and forms of communication are needed. Moreover, the "myth of happy motherhood" is common, influenced by sociocultural aspects. This myth may become unreachable and a crisis may result from this expectation, as the ideal social value is not achieved. If motherhood is culturally associated with well-being and achieving, when the son or daughter is not compatible with the one desired by the parents, as is the case of children born with a vulnerability, this condition does not only change the psychophysiological functioning of the mother and her quality of life, but can also result in negative consequences for the whole family. Parenthood is a relational experience of profound psychological meaning, experienced in family relationships, which are transformed over the entire life and that are restructured with the normal cycles of family development and, occasionally, by unforeseen events (Cerveny & Berthoud, 1997).

The arrival of a child with ASD can be considered an unexpected contingency at any stage that the family is going through, because these are serious psychiatric illnesses, which require special needs and require much understanding and patience due to the peculiarity of the symptoms. Given this reality, some authors have reported that mothers of children with disabilities tend to depression, which may be associated with hopelessnessand worsened quality of life. This is also observed in the fathers and siblings of individuals with ASDs, with the degree of symptoms reflecting the severity of the autism of the affected relative (Orsmond et al., 2009; Orsmond & Seltzer, 2009). Carter and collaborators (2009) studied stability and individual change in depressive symptoms among mothers raising young children with ASD. They observed that child problem behaviors and delayed competence, maternal anxiety symptoms and angry/hostile mood, low parenting efficacy and social supports, and coping styles were associated with depression severity. Only maternal anxiety and parenting efficacy predicted individual change. Many mothers do not appear to adapt, supporting the need for early intervention for maternal well-being.

In particular, mothers experience the reality of having an autistic child permeated by feelings of nullity, loneliness and solitude. They also stop living their daily lives to live the everyday life of the child. Brothers and sisters have more stressful conditions of life, which include early responsibilities, anxiety and feelings of inferiority (Benderix & Sivberg, 2007).

Pearson et al. (2006) found that autistic individuals have more symptoms of depression, withdrawal from social life, atypical behaviour and immature social skills. Besides, they are at particularly high risk of comorbidities involving emotional and behaviour disorders, with direct consequences on their family. Family members have to adapt to a reality that, in addition to being new, is very heterogeneous, complex and difficult and that can result in conflicts that require intervention (Kelly et al., 2008).The disease eventually becomes the focus and other problems become unimportant; family members live only the disease and end up getting sick too (Balieiro & Cerveny, 2004).

What is observed in practice is that when a child is diagnosed with ASD, parents experience a variety of very complicated feelings that are often unrelated to interventions involving the child, but related to the parents particular vision of the world(Wachtel & Carter, 2008). After all, few other diseases can pose such a great threat to the family as these do, because autism is still seen as an intense "stressor" (Woodgate et al.,2008; King et al., 2006). But when, for example, a better relationship is established between the mother and child, the autistic symptoms may reduce (Smith et al., 2008).

As ASDs are related to a great need for care that directly affects the development not only of the individual but also of their families, the resources available to families must be evaluated very well (Montalbano &Roccella, 2009; Montes & Halterman , 2008). It is important to strengthen social networks and the availability of resources such as specialized schools, stimulation therapy clinics and family psychotherapy(Smith & Elder, 2010; Cahill & Glidden, 1996). Family support is associated with increased optimism that, in turn, predict higher levels of positive feelings. Even the child psychiatrist should be encouraged to participate in the social support network of parents, helping them on the long journey of raising their children (Wachtel & Carter, 2008).

The paediatrician's role is crucial, because with more frequent contact with the child and the bond of trust with the family, the doctor will able to detect symptoms early and to guide the investigation and treatment. Most important, according to De Ocampo and Jacobs (2006), is to establish close cooperation and communication between the family and all the experts who care for the child.

There are many gaps in the scientific knowledge which justifies the need to define future research on families of children with these diseases. Health professionals must strive to study them and create effective support strategies.

5. Genetic counselor and counselee: a model and an example of case

Genetic counseling, although governed by traditional guidelines that recommend certain actions, phases and intentions, varies much in the way it is developed, from centre to centre, region to region and from country to country. Not only the emphasis on some particular goal may vary but the composition of the team and the different forms of participation of each of its members may change.

Many kinds of questions can be used in different ways to increase the understanding, respect and empathy on both sides, counselor and counselees. The counselor is part of the system in which he acts and his personality is a determinant of how the process will be conducted within the basic goals of genetic counseling. Some counselors are more paternalistic (I suffer with you and if I could do anything for the situation to be different ...), and some are more authoritarian (You have to understand that I am experienced in this matter and definitely can help you…). There are also the many peculiarities of each team; never will the counseling given by one counselor in one situation be the same as that given by another. Also, counseling performed by one team for one family with a particular type of problem will not be identical to that for another family with exactly the same problem. The process is so dynamic that it cannot be predicted.

We will briefly describe a model of genetic counseling which occurred in a community genetic service of a low-income country (Brazil). It involves a context characterized by certain cultural, legal and religious limitations such as the cultural fear of genetic disorders due to stigma and legal restrictions in respect to selective abortion, among others. The service in questionis located in a referral centre for health in a city of the most developed state of the country (São Paulo). It has an interdisciplinary team comprised of three counselors, three psychologists, physicians of different specialties, a social worker and two nurses. One of its peculiarities is that the genetic counselor and psychologist work together during counseling sessions of families, in a transdisciplinary way.

Briefly, the model can be described by the different phases through which the family passes after its arrival in the service:

- After presenting at the reception, the family is asked to stay in a waiting room. There the family is approached by a psychologist, who presents himself, establishes a rapport (contact, dialogue) and investigates the characteristics, expectations and basic needs of the family. The psychologist makes observations about the emotional state (anger, sadness, anxiety, etc.), the main coping strategies (emotional, cognitive and behavioural), psychological functions (guidance, judgement, attention, language, mood, level of understanding, etc.) and beliefs or fears. Questions such as these are used: "What is the reason for your referral to this service?", "Who referred you?" "What do you know about genetic counseling?", "How do you feel?", and "What do you expect from genetic counseling?". The family should be guided and informed on the practical, structural and dynamic operation of genetic counseling, its meaning, as well as the role of the different professionals involved. During the psychological approach a more relaxed atmosphere should be created.
- Before the counselor has contact with the family, he is informed by the psychologist on the data collected in the waiting room. The counselor has elements to promote a more focused and effective intervention, using a more targeted and personalized approach.
- On being called for counseling, the psychologist who established the initial rapport with the family in the waiting room, introduces the family members to the counselor, enters the room and participates in the genetic counseling process. Everyone sits in a circle, with a small table moved to the side, just for the counselor's note taking. The central table is considered an "obstacle" to establishing a relationship as it may suggest difference in level/hierarchical which always causes awkwardness. The psychologist accompanies the discussion, observes and only intervenes quickly and objectively on psychological aspects when requested or when he believes it is absolutely necessary. The phases of genetic counseling develop. It is up to the counselor to give psychological support inherent to the process. It is important to motivate the family to return for a follow up consultation, to perform exams, comply with treatment and to offer supports linked to the most urgent difficulties, contacting a social assistant and professionals/support institutions. When necessary, refer members of the family for a more detailed psychological assessment or for psychotherapy.
- In all consultations, the counselor and the psychologist caring for the family should be the same as the first visit and even when the process is completed the team should be available to explain future doubts that may arise through further meetings or by telephone.

All the professionals involved in the care of families of individuals with ASDs surely pass through difficult situations of intense learning that require much skill and compassion. Perhaps I can illustrate what this means using a true case.

On one day in November 2010 ... The psychologist informed the counselor that the family that she was about to meet comprised of a father, a mother, a three-year-old child with a diagnosis of autism made one week previously, and another five-year-old apparently health son. He said that the family was psychologically very weak. The mother, aged 32, expressed much sadness and spoke only when questioned. The father, 39 years old, expressed great anger, was extremely anxious and said that he did not know why they had been referred for genetic counseling, which he thought was a waste of time. Both were well educated; she is a computer engineer and university professor, and he is a judge. They had already researched on the Internet many details about the problem and were very shocked and confused. In the waiting room the psychologist explained to them about the dynamics of the process and the benefits they might obtain with the clearing up of their doubts and specific guidance. The father rejected obstinately attempts of contact and the mother reported that she was feeling very lonely. As they spoke, the eldest son always listened in silence. When asked how he felt by the psychologist, the son answered "tired".

When called and led by the psychologist to the consultation room, the counselor noted the seemingly arrogant and cold attitude of the father, who entered the room in front of his wife that was holding the hands of both children, and sat down before anyone else. The children were seated in the centre of the circle where some toys had been placed so that they could play and so they would stay there. The counselor noted the autism phenotype of the child with repetitive stereotypic movements, isolation, lack of speech, among other things, without dysmorphic signs. The mother reported that the diagnosis was made by the team of psychiatrists and neurologists that had requested exams, including biochemistry, imaging, hearing evaluation, among other tests, which were normal. The pregnancy and delivery occurred without complications. Also, there was no parental consanguinity or other risk factors involved. She said that before the completion of the diagnosis of childhood autism, other professionals had partial or wrong diagnoses, which left her very confused.

Both the father and mother started giving much information without being requested, including some technical information about autism. They started a kind of "competition", both on involving who spoke first and on the level of knowledge that each one had. Thus, the genetic counselor had an opportunity to observe and evaluate the couple's dynamics. At one point the counselor interrupted them and said, in an attempt to move on directly to emotional issues, "I am realizing how much you are frightened by the diagnosis that you received. Before I explain to you about the diagnosis, I would like to know more about your feelings. What made you so upset? Do you think it is very hard for you to talk about this now?" The couple, as they were caught by surprise, agreed to talk about it and the counselor asked the mother to speak first. Crying a lot, she reported that she was trying to understand everything that was happening and that she was not able to concentrate on her work anymore. She felt very guilty because her family was no longer the same, her eldest son was in trouble at school and that she felt very lonely. She did not know anyone with a child with the same problem and that, initially, the worst that she thought was that her son was deaf. She confessed that she always only wanted to have one child and that the second pregnancy was not planned. She had rejected the child and she felt that was being punished for this. She even felt that she was being punished too for an abortion she had as a teenager. She would like to talk to other people about their son but she had made a deal with her husband that they would not reveal the child's diagnosis to anyone; not before trying to help him to get better.

The counselor told her she had some mistaken ideas and meanings but it was good to see that she was seeking help. Those feelings, though difficult, were natural and expected, as in general, no person is prepared to have a child that is different to what they expected and very few people are ready for this possibility. The counselor continued saying that much of the information that they would receive starting from the first session, might certainly help in this difficult emotional period that the entire family was going through. The father interrupted saying "Speaking of information, I need you to tell me why my son is autistic!" The counselor felt upset with the authoritative behavior of the father and his attempts to

hide his emotions. The mother broke in with the phrase "It is impossible to live with him" to which the father replied "You cannot talk about us because is our son who needs help!"
Then the counselor told the father that she noted that he was also seeking help albeit in a different way. The counselor explained the relative proportions of autism cases in the population that might be attributable to various mechanisms of genetic transmission and that the vast majority of cases of autism remain idiopathic. The counselor asked the father how possible information about the cause of the autism of their son could help him, and why he preferred not to reveal the diagnosis to the child's relatives and friends. He replied that by discovering the cause there would certainly be drugs/specific therapies that would improve their son's condition and that people would look less and would not feel sorry for their son. The anxiety about the manifestations of the autistic child was clear. He added saying that the child was very "stubborn"; he was being seen by a speech therapist and occupational therapist and was taking psychiatric drugs and did not improve much except for being less aggressive. In a possible attempt to justify their ways, the father said many members of his family were stubborn, especially his father. He had been educated in a traditional manner. His father was very angry and never admitted that his children, all male, were weak.
The counselor noticed that the psychological defenses of father were not entirely unconscious. He was being "defensive" and his behavioral probably was related to a great sense of loss.
The counselor provided some practical explanations about autism and coping with affected children, explained the importance of knowing other families, some support institutions and the etiology of ASDs. In 90% of the cases, the etiologies of ASDs are not known.
As the parents had some technical knowledge from other sources, but did not understand it very well, the counselor re-organized the information and clearly explained it, in particular, in respect to idiopathic cases. It was explained that some more sophisticated genetic testing methods that the child had not been done, but they also had a low probability of identifying the cause of the disease. The counselor congratulated the parents because they were adhering to the proposedtreatment plan and explained the lack of specific remedies linked to a possible cause in this case and in most others. Finally, that she understood the frustration of the father, his difficulties in understanding the behaviour of his child, who was not stubborn, but he just could not "understand" what his father wanted from him.
At this point the father began to cry copiously and the psychologist intervened saying that he was among friends who wanted to help him, and that he was in the right place to express his emotions without shame or fear of being judged. The counselor asked his wife to hold the hand of her husband and in so doing the eldest son stood up and hugged his father, a move which, to everyone's surprise, was followed by the autistic son, who sat on the father's lap.
After this time, the challenges and clashes that marked the start of the session were replaced by interest to explore and discuss all information. They expressed interest in performing the tests that were missing and in doing psychotherapy. The counselor reiterated that she perceived the sense of responsibility and parental love, fundamental for the family's adjustment to the new reality. The session was adjourned with the family thanking the team for their help and patience, who thanked them for their trust. A return visit was set for 45 days.
At the next session the parents came back hand in hand, the mother was more confident and the father more pleasant. The new tests also showed normal results and the counselor

restated some information. The parents were very satisfied with psychotherapy and had chosen couples therapy. The autistic child had begun equine therapy and the parents were excited and hopeful. They said that they had organized a lunch for relatives and close friends, where they would tell their child's diagnosis and how they counted on the understanding and help of all.
The counselor and psychologist expressed their admiration and congratulated the mother and father for their initiative and expressed their satisfaction with the many positive developments. The team of professionals knew the family's feelings of love and of commitment to each other would support them through what lay ahead. At the end of the session, the psychologist could not contain himself and asked their eldest son: "And you, how are you feeling?" He just smiled and hugged his autistic brother...

6. Conclusion

ASDs have become a public health problem but there are many misunderstandings about the heritability of these disorders. The detection of genetic alterations may contribute to the diagnosis, allow an understanding of biological mechanisms involved in the pathogenesis, assist in genetic counseling of families and guide prevention and educational planning. Health care practitioners need to be able to provide information about general principles of human genetics as well as the epidemiological and molecular aspects of genetics regarding Autism Spectrum Disorders. In addition, they need to understand the limitations of genetic testing and the psychological conditions of the families. Knowledge of the genetic factors involved and of the psychological effects of these diseases is crucial for the establishment of intervention strategies that promote the bio-, psycho- and social well being of those affected and their families. Besides providing technical information necessary for the family to have a better understanding about the disease, genetic counseling can alleviate some of the common mistaken beliefs and provide support to families, assisting in the transformation and adaptation of the members. It is very important that psychoeducation programmes be created for parents, focused on handling stress and emotions, modifying false beliefs and solving the daily problems that arise from ASDs.

7. References

APA. American Psychiatric Association. (1994). DSM-IV - *Diagnostic and Statistical Manual of MentalRetardation*. American Psychiatric Association, 4th edn. Washington DC, USA.

APA. American Psychiatric Association (March 2011). DSM-V Development, In: *Proposed Revision*, 02.03.2011, Avalaible from: http:<//www.dsm5.org/Pages/Default.aspx/>.

Balieiro, C.R.B. & Cerveny, C.M.O. (2004). Família e Doença, In: *Família e...*, C. M. O. Cerveny, pp.155-166, Casa do Psicólogo, ISBN85-7396-325-5, São Paulo, Brazil.

Benderix, Y. & Sivberg, B. (2007). Siblings' Experiences of Aving a Brother or Sister with Autism and Mental Retardation: A Case Study of 14 Sibling From Five Families. *J Pediatr Nurs*, Vol.22, No.5, pp. 410-418.

Benvenuto, A.; Manzi, B.; Alessandrelli, R.; Galasso, C. & Curatolo, P. (2009). Recent Advances in the Pathogenesis of Syndromic Autisms. *Int J Pediatr*,Vol. 2009, No.ID198736, pp. 1-9.

Biederman, J.; Petty, C.R.; Fried, R.; Wozniak, J.; Micco, J.A.; Henin, A.; Doyle, R.; Joshi, G.; Galdo. M.; Kotarski, M.; Caruso, J.; Yorks, D. & Faraone, S.V. (2010). Child Behavior Checklist Clinical Scales Discriminate Referred Youth With Autism Spectrum Disorder: A Preliminary Study. *J Dev Behav Pediatr,* Vol.31, No.6, pp. 485-490.

Bremer, A.; Giacobini, M.; Eriksson, M.; Gustavsson, P.; Nordin, V.; Fernell, E.; Gillberg, C.; Nordgren, A.; Uppströmer, A.; Anderlid, B.M.; Nordenskjöld, M. & Schoumans, J. (2011). Copy Number Variation Characteristics in Subpopulations of Patients With Autism Spectrum Disorders. *Am J Med Genet B Neuropsychistr,* Vol.156, No.2, pp. 115-124.

Caglayan, A.O. (2010). Genetic Causes of Syndromic and Non-syndromic Autism. *Dev Med Child Neurol,* Vol.52, No.2, pp. 130-138.

Cahill, B.M.L. & Glidden, L.M. (1996). Influence of Child Diagnosis on Family and Parent Functioning: Down Syndrome Versus other Disabilities. *American Journal Retardation,* Vol.101, No.2, pp. 149-160.

Carter, A.S.; Martínez-Pedraza, F. de L. & Gray, S.A. (2009). Stability and Individual Change in Depressive Symptoms Among Mothers Raising Young Children with ASD: Maternal and Child Correlates. *J Clin Psychol,* Vol.65, No.12, pp.1270-1280.

Cerveny, M.O.C. & Berthoud, C.M.E. (1997). *Família e Ciclo Vital.* Casa do Psicólogo, ISCN 85-7396-002-7, São Paulo, Brazil.

Chiappedi, M.; Rossi, G.; Rossi, M.; Bejor, M. & Balottin, U. (2010). Autism and Classification Systems: a Study of 84 Children. *Ital J Pediatr,* Vol.36, pp. 10-14.

Christian, S.L.; Brune, C.W.; Sudi, J.; Kumar, R.A.; Liu, S.; Karamohamed, S.; Badner, J.A.; Matsui, S.; Conroy, J.;McQuaid, D.; Gergel, J.; Hatchwell, E.; Gilliam, T.C.; Gershon, E.S.; Nowak, N.J.; Dobyns, W.B. & Cook, E.H. Jr. (2008). Novel Submicroscopic Chromosomal Abnormalities Detected in Autism Spectrum Disorder. *Biol Psychiatry,* Vol.63, No.12, pp. 1111-1117.

Cuscó, I.; Medrano, A.; Gener, B.; Vilardell, M.; Gallastegui, F.; Villa, O.; González, E.; Rodríguez-Santiago, B.; Vilella, E.; Del Campo, M. & Pérez-Jurado, L.A. (2009). Autism-Specific Copy Number Variants Further Implicate the Phosphatidylinositol Signaling Pathway and the Glutamatergic Synapse in the Etiology of the Disorder. *Hum Mol Genet,* Vol.18, No.10, pp. 1795-1804.

De Ocampo, A.C. & Jacobs, J.M. (2006). Medical Management of Autism. *J S C Med Assoc,* Vol.102, No.8, pp. 274-276.

Gonzales, M.L. & LaSalle, J.M. (2010). The Role of MeCP2 in Brain Development and Neurodevelopmental Disorders. *Curr Psychiatry Rep,* Vol.12, No.2, pp. 127-134.

Hampson, D.R.; Adusei, D.C. & Pacey, L.K. (2011). The Neurochemical Basis for the Treatment of Autism Spectrum Disorders and Fragile X Syndrome. *Biochem Pharmacol,* Feb 17. [Epub ahead of print]

Harrington, J.W. (2010). The Actual Prevalence of Autism: Are We There Yet? *Pediatrics,* Vol.126, No.5, pp. e1257–e1258.

Homan, K.J.; Mellon, M.W.; Houlihan, D. & Katusic MZ (2011). Brief Report: Childhood Disintegrative Disorder: A Brief Examination of Eight Case Studies. *J Autism Dev Disord,* Vol.4, No.4, pp. 497-504.

Kelly, A.B.; Garnett, M.S.; Attwood, T. & Peterson, C. (2008). Autism Spectrum Symptomatology in Children: the Impact of fFamily and Peer Relationships. *J Abnorm Child Psychol,*Vol.36, No.7, pp. 1069-1081.

Kessler, S. (2001). Psychological Aspects ofGenetic Counseling. XIV. Nondirectiveness and Counseling Skills. *Genet Test*, Vol.5, No.3, pp. 187-191.

King, G.; Zwaigenbaum, L.; King, S.; Baxter, D.; Rosenbaum, P. & Bates, A. (2006). A Qualitative Investigation of Changes in Belief Systems of Families of Children With Autism or Down Syndrome. *Child Care Health Dev*, Vol.32, No.3, pp. 353-369.

Koyama, T. & Kurita, H. (2008). Cognitive Profile Difference Between Normally Intelligent Children with Asperger's Disorder and Those with Pervasive Developmental Disorder Not Otherwise Specified. *Psychiatry Clin Neurosci*, Vol.62, No.6, pp. 691-696.

Kumar, R.A. & Christian, S.L. (2009). Genetics of Autism Spectrum Disorders. *Curr Neurol Neurosci Rep*, Vol.9, No.3, pp. 188-197.

Lintas, C. & Persico, A.M. (2008). Autistic Phenotypes and Genetic Testing: State-of-the-art for the Clinical Geneticist. *J Med Genet*, Vol.46, No.1, pp. 1-8.

Liu, K.Y.; King, M. & Bearman, P.S. (2010). Social Influence and the Autism Epidemic. *AJS*, Vol.115, No.5, pp. 1387-434.

Losh, M.; Childress, D.; Lam, K. & Piven J. (2008). Defining Key Features of the Broad Autism Phenotype: a Comparison Across Parents of Multiple- and Single-Incidence Autism Families. *Am J Med Genet B Neuropsychiatr Genet*, Vol.147B, No.4, pp. 424-433.

Maenner, M.J. & Durkin, M.S.(2010). Trends in the Prevalence of Autism on the Basis of Special Education Data, *Pediatrics*, Vol.126, No.5, pp. e1018–e1025.

Marshall, C.R.; Noor, A.; Vincent, J.B.; Lionel, A.C.; Feuk, L.;Skaug, J.; Shago, M.; Moessner, R.; Pinto, D.; Ren, Y.; Thiruvahindrapduram, B.; Fiebig, A.; Schreiber, S.; Friedman, J.; Ketelaars, C.E.; Vos, Y.J.; Ficicioglu, C.; Kirkpatrick, S.; Nicolson, R.; Sloman, L.; Summers, A.; Gibbons, C.A.; Teebi, A.; Chitayat, D.; Weksberg, R.; Thompson, A.; Vardy, C.; Crosbie, V.; Luscombe, S.; Baatjes, R.; Zwaigenbaum, L.; Roberts, W.; Fernandez, B.; Szatmari, P. & Scherer, S.W. (2008). Structural Variation of Chromosomes in Autism Spectrum Disorder. *Am J Hum Genet*, Vol.82, No.2, pp. 477-88.

Marteleto, M.R.; Lima e Menezes, C.G.; Tamanaha, A.C.; Chiari, B.M. & Perissinoto, J. (2008). Administration of the Autism Behavior Checklist: Agreement Between Parents and Professionals' Observations in Two Intervention Contexts.*Rev Bras Psiquiatr*, Vol.30, No.3, pp.203-208.

Miles, J.H.; McCathren, R.B.; Stichter, J. & Shinawi, M. (April, 2010). *Autism Spectrum Disorders*, BookShelf, Retrieved from
<http://www.ncbi.nlm.nih.gov/books/NBK1442/>

Montalbano, R. & Roccella M. (2009). The Quality of Life of Children with Pervasive Developmental Disorders. *Minerva Pediatr*, Vol.61, No.4, pp. 361-370.

Montes, G. & Halterman, J.S. (2008). Child Care Problems and Employment Among Families with Preschool-Aged Children With Autism in the United States. *Pediatrics*, Vol.122, No.1, pp. e202-e208.

Moss, J. & Howlin, P. (2009). Autism Spectrum Disorders in Genetic Syndromes: Implications for Diagnosis, Intervention and Understanding the Wider Autism Spectrum Disorder Population. ,Vol.53, No.10, pp. 852-873.

Nassar, N.; Dixon, G.; Bourke, J.; Bower, C.; Glasson, E.; de Klerk, N. & Leonard, H. (2009). Autism Spectrum Disorders in Young Children: Effect of Changes in Diagnostic Practices. *Int J Epidemiol*, Vol.38, No.5, pp. 1245-1254.

Nissenkorn, A.; Gak, E.; Vecsler, M.; Reznik, H.; Menascu, S. & Ben Zeev, B. (2010).Epilepsy in RettSyndrome -The experience of a National Rett Center. *Epilepsia,* Vol.51, No.7, pp. 1252-1258.

Orsmond, G.I.; Kuo, H.Y. & Seltzer, M.M. (2009). Siblings of Individuals with an Autism Spectrum Disorder: Sibling Relationships and Wellbeing in Adolescence and Adulthood.*Autism*, Vol.13, No.1, pp. 59-80.

Orsmond, G.I. & Seltzer, M.M. (2009). Adolescent Siblings of Individuals with an Autism Spectrum Disorder: Testing a Diathesis-Stress Model of Sibling Well-Being. *J Autism Dev Disord*, Vol.39, No.7, pp. 1053-1065.

Pearson, D.A.; Loveland, K.A.; Lachar, D.; Lane, D.M.; Reddoch, S.L.; Mansour, R. & Cleveland, L.A. (2006).A Comparison of Behavioral and Emocional in Children and Adolescents With Autistic Disorder and PDD-NOS. *Child Neuropsychol*, Vol.12, No.4-5, pp. 321-333.

Peters, K.F. & Petrill, S.A. (2011).Development of a Scale to Assess the Background, Needs, and Expectations of Genetic Counseling Clients. *Am J Med Genet A*. Feb 23. doi: 10.1002/ajmg.a.33610. [Epub ahead of print]

Ratajczak, H.V. (2011). Theoretical Aspects of Autism: Causes - a Review. *J Immunotoxicol*, Vol.8, No.1, pp. 68-79.

Schmidt, G.L.; Kimel, L.K.; Winterrowd, E.; Pennington, B.F.; Hepburn, S.L. & Rojas, D.C. (2008). Impairments in Phonological Processing and Nonverbal Intellectual Function in Parents of Children With Autism. *J Clin Exp Neuropsychol*, Vol.30, No.5, pp. 557-567.

Selkirk, C.G.; McCarthy Veach, P.; Lian, F.; Schimmenti, L. & LeRoy, B.S. (2009). Parents' Perceptions of Autism Spectrum Disorder Etiology and Recurrence Risk and Effects of their Perceptions on Family Planning: Recommendations for Genetic Counselors. *J Genet Couns*, Vol.18, No.5, pp. 507-519.

Shen, Y.; Dies, K.A.; Holm, I.A.; Bridgemohan, C.; Sobeih, M.M.; Caronna, E.B. Miller, K.J.; Frazier, J.A.; Silverstein, I.; Picker, J.; Weissman, L.; Raffalli, P.; Jeste, S.; Demmer, L.A.; Peters, H.K.; Brewster, S.J.; Kowalczyk, S.J.; Rosen-Sheidley, B.; McGowan, C.; Duda, A.W. 3rd; Lincoln, S.A.; Lowe, K.R.; Schonwald, A.; Robbins, M.; Hisama, F.; Wolff, R.; Becker, R.; Nasir, R.; Urion, D.K.; Milunsky, J.M.; Rappaport, L.; Gusella, J.F.; Walsh, C.A.; Wu, B.L. & Miller, D.T. Autism Consortium Clinical Genetics/DNA Diagnostics Collaboration. (2010). Clinical Genetic Testing for Patients With Autism Spectrum Disorders. *Pediatrics*, Vol.125, No.4, pp. e727-735.

Smith, L.E.; Greenberg, J.S.; Seltzer, M.M. & Hong, J. (2008). Symptoms and Behavior Problems of Adolescents and Adults With Autism: Effects of Mother-Child Relationship Quality, Warmth, and Praise. *Am J Ment Retard*, Vol.113, No.5, pp. 387-402.

Smith, L.O. & Elder, J.H. (2010). Siblings and Family Environments of Persons With Autism SpectrumDisorder: a Review of the Literature. *J Child Adolesc Psychiatr Nurs*, Vol.23, No.3, pp. 189-195.

Snow, A.V. & Lecavalier, L. (2011). Comparing Autism, PDD-NOS, and Other Developmental Disabilities on Parent-Reported Behavior Problems: Little Evidence for ASD Subtype Validity. *J Autism Dev Disord*, Vol.41, No.3, pp. 302-310.

Sykes, N.H. & Lamb, J.A. (2007). Autism: the Quest for the Genes. *Expert Rev Mol Med*, Vol.9, No.24. pp. 1-15.

Temudo, T.; Santos, M.; Ramos, E.; Dias, K.; Vieira, J.P.; Moreira, A.; Calado, E.; Carrilho, I.; Oliveira, G.; Levy, A.; Barbot, C.; Fonseca, M.; Cabral, A.; Cabral, P.; Monteiro, J.; Borges, L.; Gomes, R.; Mira, G.; Pereira, S.A.; Santos, M.; Fernandes, A.; Epplen, J.T.; Sequeiros, J. & Maciel, P. (2011). Rett syndrome With and Without Detected MECP2 Mutations: An Attempt to Redefine Phenotypes. *Brain Dev,* Vol.33, No.1, pp. 69-76.

Tuchman, R.; Cuccaro, M. & Alessandri, M. (2010). Autism and Epilepsy: Historical Perspective. *Brain Dev*, Vol.32, No.9, pp. 709-718.

Vorstman, J.A.; Staal, W.G.; van Daalen, E.; van Engeland, H.; Hochstenbach, P.F. & Franke, L. (2006). Identification of Novel Autism Candidate Regions Through Analysis of Reported Cytogenetic Abnormalities Associated With Autism. *Mol Psychiatry,* Vol.11, No.1, pp. 18-28.

Wachtel, K. & Carter, A.S. (2008).Reaction to Diagnosis and Parenting Styles Among Mothers of Young Children with ASDs. *Autism*, Vol.12, No.5, pp. 575-594.

Wassink, T.H.; Losh, M.; Piven, J.; Sheffield, V.C.; Ashley, E.; Westin, E.R. & Patil, S.R. (2007). Systematic for Subtelomeric Anomalies in a Clinical Sample of Autism. *J Autism Dev Disord*,Vol.37, No.4, pp. 703-708.

Weil, J. (2000). *Psychosocial Genetic Counseling*. Oxford, ISBN 978-019-512066-0, New York, USA.

WHO.World Health Organization (1998). *Proposed International Guidelines on Ethical Issues in Medical Genetics and Genetic Services: Report of WHO Meeting on Ethical Issues in Medical Genetics.* World Health Organization, Geneva.

WHO (2010). *Community Genetics Services: Report of WHO Consultation on Community Genetics in Low- and Middle-Income Countries.* World Health Organization, Geneva.

Witwer, A.N. & Lecavalier, L. (2008). Examining the Validity of Autism Spectrum Disorder Subtypes. *J Autism Dev Disord*, Vol.38, No.9, pp. 1611-1624.

Wood, J.J.; Drahota, A.; Sze, K.; Van Dyke, M.; Decker, K.; Fujii, C.; Bahng, C.; Renno, P.; Hwang, W.C. & Spiker, M. (2009). Brief Report: Effects of Cognitive Behavioral Therapy on Parent-Reported Autism Symptoms in School-Age Children With High-Functioning Autism.*J Autism Dev Disord,* Vol.39, No.11, pp. 1608-1612.

Woodgate, R.L.; Ateah, C. &Secco, L. (2008). Living in a World of Our Own: the Experience of Parents Who Have a Child With Autism. *Qual Health Res*, Vol.18, No.8, pp. 1075-1083.

Zhang, X.; Lv, C.C.; Tian, J.; Miao, R.J.; Xi, W.; Hertz-Picciotto, I. & Qi, L. (2010). Prenatal and Perinatal Risk Factors for Autism in China. *J Autism Dev Disor,* Vol.40, No.11, pp. 1311-21.

Zahir, F.R. & Brown, C.J. (2011). Epigenetic Impacts on Neurodevelopment: Pathophysiological Mechanisms and Genetic Modes of Action. *Pediatr Res*, Feb 2. [Epub ahead of print]